ALGEBRA EXAMPLES

CONIC 4 ELLIPSES

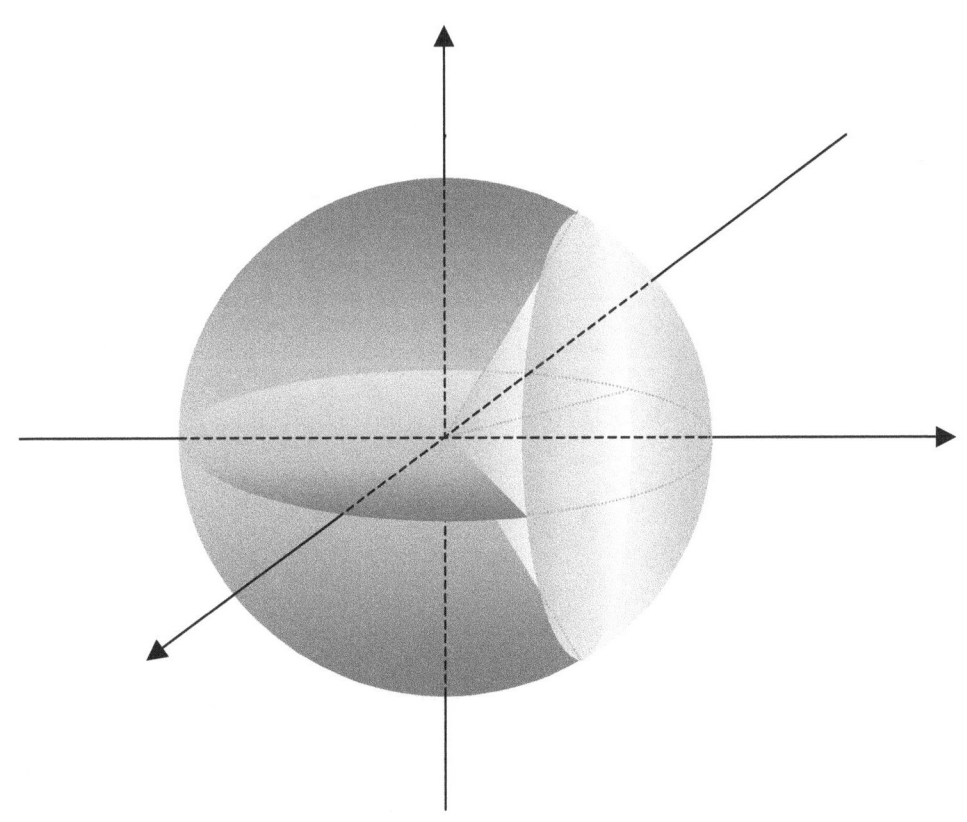

Seong R. Kim

Dear students:

Students need the best teacher, so you need examples, because examples are the best teacher. All the examples here are fully worked, and explain **how** the basic and essential tools in math are made, together with **what** they are, **how** they work, and **how** to work with them. Such tools include numbers, formulas, identities, equations, laws, etc.

Examples here begin with easy ones, of course. Covering every meter and yard properly, we can cover thousands of miles and kilometers. And it is particularly the case in math.

Of those examples therefore, some might even look too easy for you. It's not that easy though, to come up with those examples. Anyways, the bigger and the taller the tree, the deeper and the stronger the root.

Doing math, we work with ideas and run ideas, because every thing in math is an idea. A number is an idea, for instance, and the same is true for a line or circle, too. And putting ideas together, we build another, which becomes the base or an element of another, and each is connected. And that's the way your math grows. So you get to build a circuit, and sometimes, need to fill the gap or repair the circuit so that you get the sense of it.

So your calculation runs properly, and you get the problem solved.

The examples have been made and arranged so that they get tougher (or sometimes easier for some reason) as you proceed with them. In particular, similar examples with some variations are strategically repeated so that you can get the ideas or the tools tricky or complicated, and can get them mastered.

This book is however, nothing but a bunch of examples until you get it powered. How then, to get it powered, and make it run and work for you?

Just read it, and then, do each example in writing. And it is important to note that you do it in **your** writing. Just watching someone doing it, you just only feel that you can do it. If you do it, you can do it, but if you don't, we can hardly. It's a cliché, of course, but is always true that knowing is one thing and doing is another.

I've been helping students grow, take care of, and run their own math. The area covers algebra and geometry for high school or college students, and is especially for equations (for unknowns or curves), functions, and their graphs, which are the basic elements in calculus, which's been the core of my interest from my early age in high school.

Of my students, some are quite poor in math, and thus, are afraid of or hate math, some require special education because of exceptional intelligence, some are smart enough, some are naïve and diligent, some are clever but lazy, and most behave in general. All the students are badly after though, one thing in common: a strong and secure math skill. It is of course, the prime objective of my work, and I'm always happy to and eager to help them achieve it. The problem was however, that many of them wanted it to be purchased. And the question is, can we buy it?

We can buy the means, of course. And a solid math skill is feasible, too. We know however, we can't buy love, and the same is true for the math skill, too. It's not what we can buy or sell, and not what we can give or take. It is however, what we can grow, and need to grow. Your math grows as much as you grow and take care of it. So does mine.

What math then, do students most often do or use in high schools or colleges?

It is algebra and geometry. What algebra though?

Elementary algebra, of course
Doing the algebra, we work with numbers (many in kinds), constants, variables, ratios, rates, expressions, equations, inequalities, functions, identities, formulas, laws, etc., together with signs and symbols. And if we want to do algebra properly, we want to know their natures and how they mingle with each other.

So studying math ideas or tools, you want to know **what** they are, **how** they work, and **how** to work with them or **what** to do with them.   What then, about the geometry?

Basically, the geometry has much to do with shapes, positions, and angles. The shapes begin with triangles and circles, and move on to rectangles, squares, parallelograms or rhombuses, trapezoids, tetragons, other polygons, polyhedrons, etc.

Doing the geometry, too, though, we need to do the algebra stated above. So it is analytic geometry, often called coordinate geometry, too. And doing it, we can specify positions using coordinates. So in the geometry, basically, we work with graphs. Putting a math idea in a graph, we can not only effectively think about it but actually see it, too, and therefore, can efficiently work with it. What idea then, is it?

The idea begins with a point, line, parabola, circle, ellipse, and hyperbola, called a conic section or basic curve, and then, moves on to other curves, planes, surfaces, volumes, and other objects in various dimensional spaces, together with vectors.

And using an angle, we can specify an amount of turn or change in direction.

So learning, using, or applying those ideas or math tools, we get to solve problems.

And this book can help. It can help learn them, and use them so that you can navigate to find solutions to problems. And in particular, it can help come up with answers to those **what**s and **how**s stated above. So it can help you grow and run your own math, and thus, can help achieve your solid math skill.

It is however, not a magic book giving you a math skill of high caliber overnight. And it can have many mistakes, too. There is no magic, and math is full of facts and ideas. And it is after all, not me and not your teacher but you who put together some of those facts and ideas, and understand it. Putting facts and ideas together, understanding it, and taking care of what you have learned, you grow your math. And this book can help.

This is a book of examples designed to help you grow your math, and assumes that you are a real beginner. This book requires though, time and effort, the amount of which need to be substantial, too, but will be worth it. That's because you want a substantial achievement, and will get it. And probably, you will get to see this book helping you get there much faster than expected. And then, you will get to see the way math runs.

In math, everything is an idea. So is a problem. And solving it, we put it many different ways. For instance, while expanding or reducing it, or modifying or converting it, we keep searching for the solution, approaching the solution, and eventually, can get there. So don't look for the solution outside the problem. The solution is inside the problem if the problem is properly made.

If it is not, no solution is the solution. And in fact, it is often the case a problem itself is the solution. We can put a problem in many different ways, and eventually, can end up with the solution.   How come then, is the solution no other than the problem?

For instance, the solution to $3232 \div 101$ is 32.   And we can put it this way:

$$3232 \div 101 = \frac{3232}{101} = \frac{32 \times 101}{101} = \frac{32}{1} = 32 \implies 3232 \div 101 = 32.$$

And we can get this, too: $32 \implies 3232 \div 101$.   How?

$$32 = \frac{32}{1} = \frac{32 \times 101}{101} = \frac{3232}{101} = 3232/101 = 3232 \div 101.   \text{Too easy?}$$

For another instance, the solution to $ax^2 + bx + c = 0$ is: $x = \frac{-b \pm \sqrt{b^2 - 4ac}}{2a}$, which is called the quadratic formula.   How come then, is the solution no other than the problem?

We can put it this way:

$$x = \frac{-b \pm \sqrt{b^2 - 4ac}}{2a} \implies 2ax = -b \pm \sqrt{b^2 - 4ac} \implies 2ax + b = \pm\sqrt{b^2 - 4ac}$$

$$\implies (2ax + b)^2 = b^2 - 4ac \implies 4a^2x^2 + 4abx + b^2 = b^2 - 4ac$$

$$\implies 4a^2x^2 + 4abx = -4ac \implies ax^2 + bx = -c \implies ax^2 + bx + c = 0.$$

And we can get this, too: $ax^2 + bx + c = 0 \implies x = \frac{-b \pm \sqrt{b^2 - 4ac}}{2a}$.   How?

$$ax^2 + bx + c = a(x^2 + \tfrac{b}{a}x) + c = a(x^2 + \tfrac{b}{a}x + \tfrac{b^2}{4a^2} - \tfrac{b^2}{4a^2}) + c = a(x^2 + \tfrac{b}{a}x + \tfrac{b^2}{4a^2}) - \tfrac{b^2}{4a} + c$$

$$= a(x + \tfrac{b}{2a})^2 - \tfrac{b^2 - 4ac}{4a} = 0 \implies a(x + \tfrac{b}{2a})^2 = \tfrac{b^2 - 4ac}{4a} \implies (x + \tfrac{b}{2a})^2 = \tfrac{b^2 - 4ac}{4a^2} \implies x + \tfrac{b}{2a} = \pm\sqrt{\tfrac{b^2 - 4ac}{4a^2}}$$

$$\implies x = -\tfrac{b}{2a} \pm \tfrac{\sqrt{b^2 - 4ac}}{2a} = \tfrac{-b \pm \sqrt{b^2 - 4ac}}{2a} \implies x = \tfrac{-b \pm \sqrt{b^2 - 4ac}}{2a}.$$

And we call the set of processes above, algebra.

So if a problem is well defined, that is, if it makes sense, we should be able to get it solved the way below:

A problem ⇒ … ⇒ … ⇒ the solution, and thus: **the problem ⇒ the solution**.

So solving a problem, we put it many different ways so that we can get to the solution.

And that's the way, math runs.

May your math run very well.

Seong R. Kim

B.S. Math. Michigan Tech. Univ.   M.S. Math. Rensselaer Polytechnic Institute

Notes:

This book is one of five books about some basics in elementary algebra, and covers equations often used in high schools and colleges or universities. And the equations are for curves called conic sections, often just called conics. There are five kinds in conics. And of the five, one is covered briefly here in this book, and is for ellipses.

So this book covers equations indicating ellipses, that is, equations for ellipses. And this book explains what such an equation is about, how it gets made, what it does or how it behaves, and what we can do with it or how to use it.   What then, is it for?

An ellipse is an idea in math, so it's a math idea, and is a tool in math. So it's a math tool. And we use it, solving problems, of course. So students need to get the idea.

And thus, this book helps you get the idea of an ellipse, that is, the concept of a math object called an ellipse, and see how to use it, because the book explains what it is and how it works, along with those stated above so that you can develop your own idea to make use of it, solving problems, of course.   What then, about the other conics?

They are covered in their individual books, too, which are as follows:

Algebra Examples Conics 1 Lines, which covers therefore, equations for lines, often called linear equations or equations of degree 1.

Algebra Examples Conics 2 Parabolas, which covers thus, equations for parabolas, often called quadratic equations or equations of degree 2.

Algebra Examples Conics 3 Circles, which is about equations for circles.

Algebra Examples Conics 5 Hyperbolas, explaining those for hyperbolas.

And each book is designed for those students who want to study calculus, want to major in science or engineering, or want to take IB courses in math, so each book covers materials in each category in such a depth and extent. And if you don't need that much, it will be sufficient to study the sections and the sets of examples bulleted in the table of contents.

Either way, the books will help you grab math ideas often used in real life as well as in math courses. The ideas are about lines, parabolas, circles, ellipses, hyperbolas, and their equations, so you will get to see what those curves and equations are about, how they work, and how to use them, and develop your own idea to make use of those, solving problems, of course.

In short, the books help you develop and grow your own idea to make use of math ideas, providing examples, showing all the steps and the ideas behind them, and explaining what the math ideas are about.

Contents

$$(x + y)^2 = x^2 + 2xy + y^2.$$

$$(x + y)^3 = x^3 + 3x^2y + 3xy^2 + y^3.$$

$$(x + y)(x - y) = x^2 - y^2.$$

$$(x + y)(x^2 - xy + y^2) = x^3 + y^3.$$

$$(x^2 + xy + y^2)(x^2 - xy + y^2) = x^4 + x^2y^2 + y^4.$$

$$(x + a)(x + b) = x^2 + (a + b)x + ab.$$

$$(ax + b)(cx + d) = acx^2 + (ad + bc)x + bd.$$

$$(x + a)(x + b)(x + c) = x^3 + (a + b + c)x^2 + (ac + bc + ca)x + abc.$$

$$(a + b + c)^2 = a^2 + b^2 + c^2 + 2(ab + bc + ca).$$

$$(a + b + c)(a^2 + b^2 + c^2 - ab - bc - ca) = a^3 + b^3 + c^3 - 3abc.$$

Suppose both a and $b \neq 0$, and both m and n are integers. Then, we get:

0. $a^m a^n = a^{m+n}$ **1.** $a^m / a^n = \dfrac{a^m}{a^n} = a^{m-n}$ **2.** $(a^m)^n = a^{mn}$

3. $(ab)^n = a^n b^n$ **4.** $(a/b)^n = \left(\dfrac{a}{b}\right)^n = a^n / b^n = \dfrac{a^n}{b^n}$

Suppose both a and $b > 0$, and m and n both are integers nonzero. Then, we get:

0.1. $a^{\frac{1}{n}} b^{\frac{1}{n}} = (ab)^{\frac{1}{n}}$.   **1.1.** $\dfrac{a^{\frac{1}{n}}}{b^{\frac{1}{n}}} = \left(\dfrac{a}{b}\right)^{\frac{1}{n}}$.   **2.1.** $(a^{\frac{1}{n}})^m = (a^m)^{\frac{1}{n}}$.

3.1. $(a^{\frac{1}{n}})^{\frac{1}{m}} = a^{\frac{1}{mn}} = (a^{\frac{1}{m}})^{\frac{1}{n}}$.   **3.2.** $(a^{mp})^{\frac{1}{np}} = (a^m)^{\frac{1}{n}}$, where p is a nonzero integer.

1. Suppose M, N, and $b > 0$, but $b \neq 1$, and we have: $A = \log_b M$, and $B = \log_b N$. Then, we get: $A - B = \log_b M - \log_b N = \log_b \frac{M}{N}$.

2. Suppose that M and $b > 0$, but $b \neq 1$, and that we have: $E = \log_b M$. Then, we get: $PE = P \log_b M = \log_b M^P$.

3. Suppose that a, b, C, and $D > 0$, but a and $b \neq 1$, and that we have: $\log_a C = \log_b D$. Then, we get: $\log_a C = \log_b D = \log_{ab} CD$.

4. Suppose that a, b, C, and $D > 0$, but a and $b \neq 1$, and that we have: $\log_a C = \log_b D$. Then, we get: $\log_a C = \log_b D = \log_{\frac{a}{b}} \frac{C}{D} = \log_{\frac{b}{a}} \frac{D}{C}$.

5. $\log_b b = 1$, and $\log_b 1 = 0$.   **6.** $\log_b A = \dfrac{\log_c A}{\log_c b}$.

7. $\log_b A = \dfrac{1}{\log_A b}$.

Note:

The drawings or graphs in this book are not exact, and are approximate or conceptual ones.

$\in$	"$a \in B$" means that a belongs to B. "p, q, and $r \in W$" means that p, q, and r belong to W.
$\Rightarrow$	"$A \Rightarrow B$." means that A implies B.
$\equiv$	$A \equiv B$ means that A and B are identical to each other.
$\neq$	$A \neq B$ means that A is not equal to B.
$\lvert A \rvert$	The magnitude of A. For instance, $\lvert -1 \rvert = \lvert 1 \rvert = 1$.
$\therefore$	Therefore
$\Leftrightarrow$	"$A \Leftrightarrow B$" means "If A then B." and "If B then A." We can read $A \Leftrightarrow B$ as "A if and only if B." In such a case, we can say that $A = B$.
Δx and Δy	Suppose that (x_1, y_1) and (x_2, y_2) are two points in the x-y plane. Then, we get either of the two below. $\Delta x = x_2 - x_1$, and $\Delta y = y_2 - y_1$. $\Delta x = x_1 - x_2$, and $\Delta y = y_1 - y_2$.

Distance Formula

Suppose that d is the distance between two points (x_1, y_1) and (x_2, y_2) in the x-y plane. Then, we get $d^2 = (\Delta x)^2 + (\Delta y)^2$.

₀. **What is an ellipse?**

An ellipse is in a plane, and is a set of points, from each of which, <u>the sum of the two distances to two particular points is constant</u>, that is, the same. And the two particular points are called the foci of the ellipse.

An ellipse is one of the basic curves called conic sections, often called conics, for short.

So it is a conic, and we call it an oval, too, since it looks like an egg. And also, it can be called a long circle, too.   What do we mean by a long circle?

An ellipse can be quite close to a circle, but can be quite different from a circle, too. So to begin with, comparing an ellipse with a circle, we can say that a circle can be an incircle or a circumcircle of a square, and that an ellipse can be inscribed in or can be circumscribed about a rectangle.

An incircle of a square is tangent to all the four sides of the square, and a circumcircle of a square is passing through all the four vertices of the square. So an ellipse inscribed in a rectangle is tangent to all the four sides of the rectangle, and an ellipse circumscribed about a rectangle is passing through all the four vertices of the rectangle.

circles and a square   ellipses and a rectangle

Fig. 0

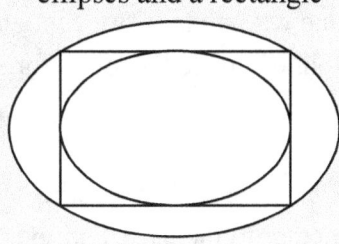

Next, a circle is a simply closed curved-line segment that is symmetric and has infinitely many axes of symmetry, each of which is a diameter.

An ellipse, too, is a simply closed curved-line segment that is symmetric.

It has however, only **two axes of symmetry**, one is called a **major axis**, and the other is called a **minor axis**, called a conjugate axis, too. And we call the two axes of symmetry **main axes**, too, and can put the main axes in an ellipse the way below:

Fig. 1

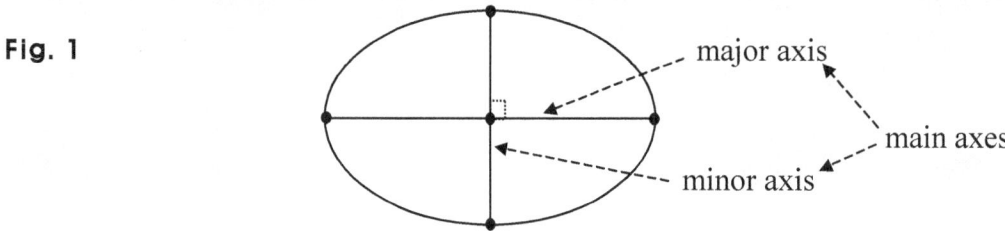

As shown above, the two main axes are perpendicular to each other.

And we can notice that the major axis connects two points the farthest away from the center of an ellipse, and that the minor axis connects two points the closest to the center.

In other words, the major axis connects two antipodal points the farthest away from each other, and the minor axis connects two antipodal points the closest to each other.

And of course, the two main axes meet at the center.

How then, can we get an ellipse?

If cutting a right cone with a plane, we can get a cross section called an ellipse. So it's called a conic section, just called a conic, for short.   What is a right cone though?

If a cone is a right cone, the line connecting the vertex of the right cone and the center of the base is perpendicular to the base, and thus, makes a right angle (90°) with the base.

And the line stated above is called the axis of the cone. And in the study of conics, such a right cone is said to be made or generated the way as follows:

First, take a line as the axis of the cone to be made.
Next, make another line meet the axis of the cone at a point at an angle γ, which is an acute angle, that is, $0 < \gamma < 90^\circ$.

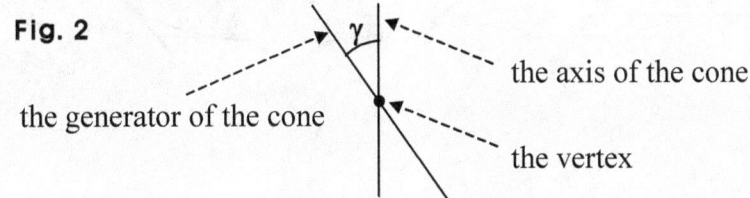

Fig. 2

the axis of the cone

the generator of the cone

the vertex

Then, we call the other line the *generator of the cone*, and the point where the other line, that is, the generator of the cone meets the axis of the cone is the vertex of the cone.

So the vertex of the cone is the point where the generator meets the axis of the cone.

And in the figure above, the angle γ is called the *generator angle*.

Then, rotating the generator around the axis fixing the generator at the vertex, we get a double-cone as shown below:

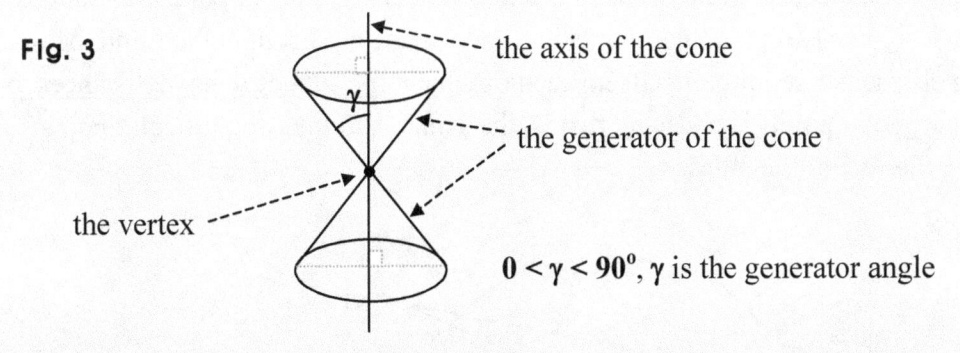

Fig. 3

the axis of the cone

the generator of the cone

the vertex

$0 < \gamma < 90^\circ$, γ is the generator angle

And note that the generator and the axis are lines, and lines have infinite lengths, so the lengths of the cones are infinite, too. So we cannot show them all. Showing thus, such a cone or a curve in math, we just show some part of it.

How then, can we get an ellipse cutting the cone with a plane?

The cross section in black below is a curve closed, and is an ellipse. So if α is the angle between the plane and the axis of the cone, and $\gamma < \alpha < 90^\circ$, and of course, if the plane does not include the vertex, the cross section is an ellipse.

Fig. 4

If however, $\alpha = \gamma$, the cross section is a parabola, and if $\alpha = 90^\circ$, the cross section is a circle.   And we can explain an ellipse the way as follows:

An ellipse is a collection of points as in the case of a circle, line, or parabola. Suppose now, we designate two particular points in the *x-y* plane as shown in the figure below. Then, an ellipse is a set of points, from each of which, <u>the sum of the two distances to the two particular points is constant,</u> that is, the same. And the two particular points are called the foci of the ellipse.

Fig. 5

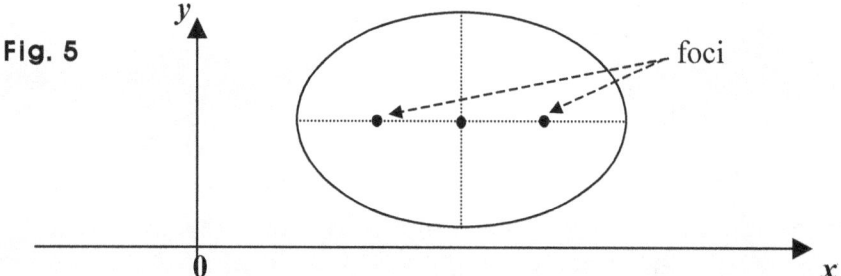

And we can also explain an ellipse the way as follows.

Suppose that a point is moving along a curve in the *x-y* plane, and that the sum of the two distances from the moving point to two particular points is constant.
Then, the two particular points are called the foci, and the curve is an ellipse.

So if a point *(x, y)* is in an ellipse, no matter where the point *(x, y)* may be, <u>the sum of the two distances from the point *(x, y)* to the two points called the foci is constant.</u>

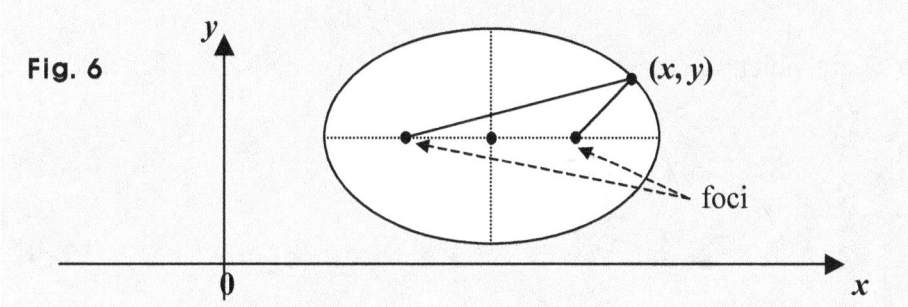

Fig. 6

And we can notice that one of the two main axes is perpendicular to the *x*-axis. In the figure above, the minor axis is perpendicular to the *x*-axis. And in that case, of course, the major axis is parallel to the *x*-axis.

It can be the case, of course, no main axis is perpendicular to the *x*-axis. Such an ellipse can be said to be skewed or tilted, and some examples can be as follows:

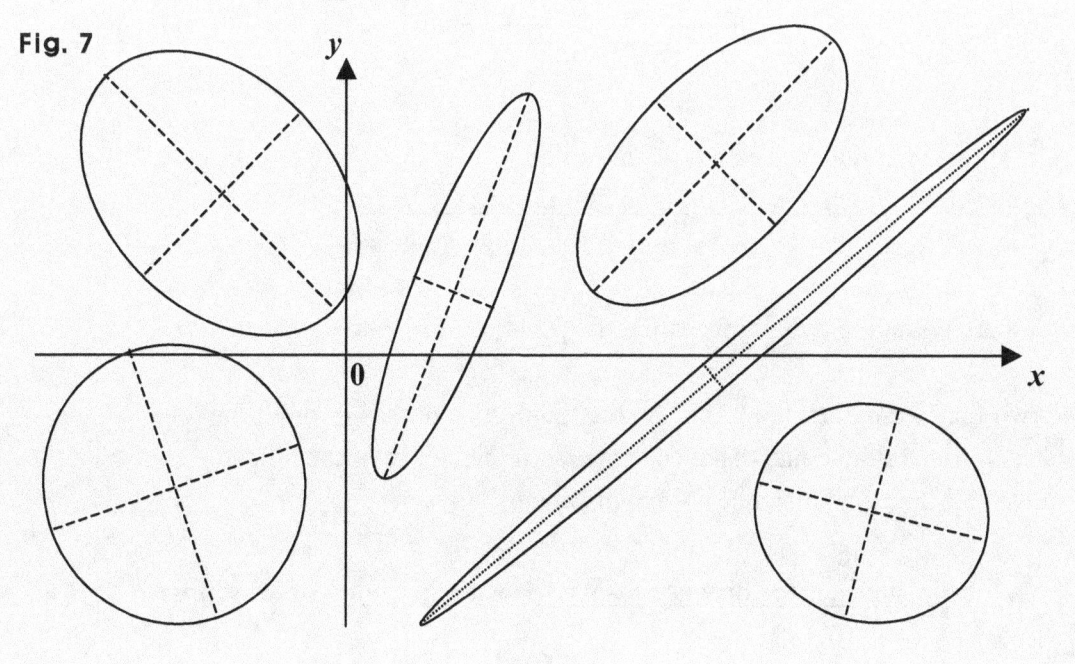

Fig. 7

(And some ellipses are quite close to circles, and some others look like bars.)

Usually though, in high school math or basic courses in college math, if we use an ellipse, one of the two main axes is <u>perpendicular to a coordinate axis</u>. So such an ellipse can be called a <u>perpendicular ellipse</u>. And in fact, <u>in this book</u>, such an ellipse is said to be perpendicular, and thus, is called a perpendicular ellipse. (Note however, in other books, it may not be called a perpendicular ellipse, and can be called differently.)

Anyway, putting a perpendicular ellipse in the *x-y* plane, we can get:

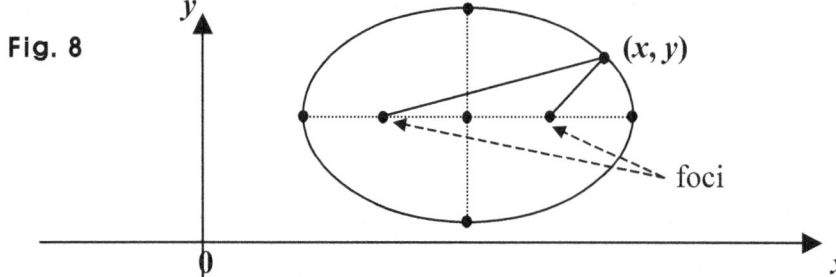

Fig. 8

And looking at the figure above, we can describe the two main axes the way below, too:

The longer of the two is called the major axis. And <u>the major</u> axis connects two points, <u>from each of which, the difference between the distances to the foci is the largest.</u>

The shorter main axis is called the minor axis. And <u>the minor</u> axis connects two points, <u>from each of which, the distances to the foci are the same.</u>

What then, about the two points called the foci?

The two foci are in the major axis. So the major axis passes through the foci.
And each focus is an equal distance away from the center of the ellipse.
And we call the equal distance a focal distance.

So <u>a focal distance is the distance from a focus to the center</u> of the ellipse.

And <u>in this book</u>, we have two kinds in ellipses perpendicular.
One is horizontal, and the other is vertical.

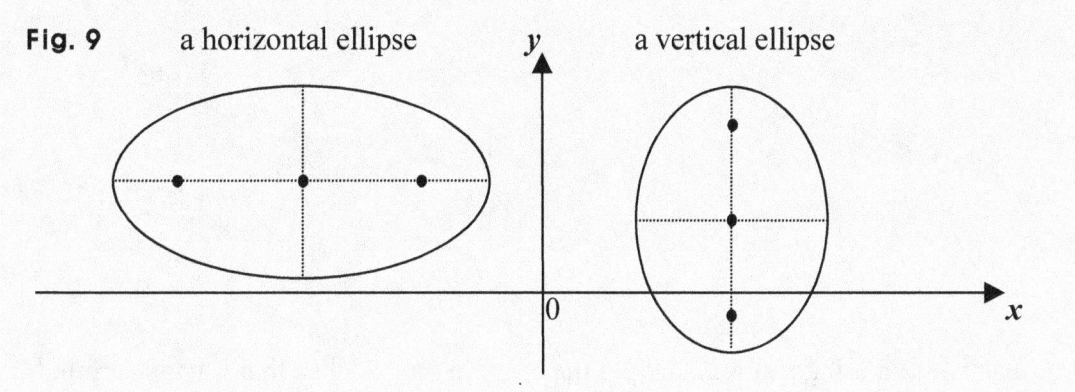

Fig. 9   a horizontal ellipse   *y*   a vertical ellipse

So in an ellipse horizontal, the major axis is horizontal, that is, parallel to the *x*-axis, and in an ellipse vertical, the major axis is vertical, that is, perpendicular to the *x*-axis.
In short, if the major is horizontal, the ellipse is horizontal, and if the major is vertical, the ellipse is vertical.

(Note that though, it is the case in this book, and <u>it may not be the case in other books</u>. That is to say that in other books, ellipses may not be classified the way above, and thus, may not be said to be horizontal or vertical.)

Either way, the foci are in the major axis, and the focal distance is the distance from a focus to the center of an ellipse.

And usually, a **half major** axis is called a **semi major** axis or a **major radius**, so a half minor axis is called a semi minor axis or a minor radius.

Where can we find an ellipse though?

We know that the earth is orbiting around the sun, and the orbit is an ellipse.
Many orbits are ellipses. Probably, every orbit is an ellipse. And also, if an object looks circular, it is quite likely to be an ellipse. How can we actually make an ellipse though?

We know the fact that an ellipse is a set of points, from each of which, the sum of the two distances to the foci is constant.   So?

So using the fact above, we can construct an ellipse the way as follows.

First, set the two foci by fastening, for instance, two nails or thumb tacks to a panel.

Fig. A

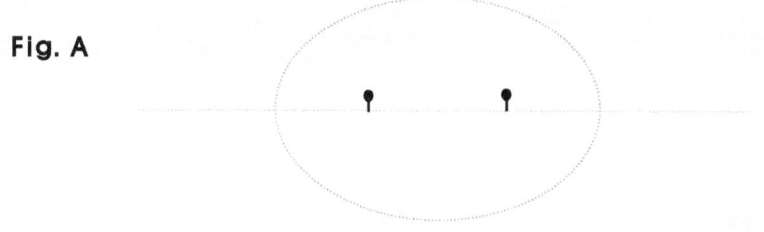

Next, make a loop using a string, and put the loop on the panel so that it wraps around the nails.

Fig. B

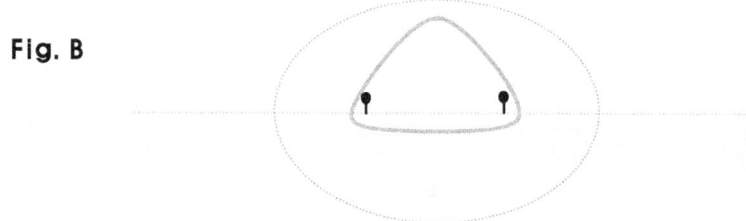

Then, while holding the loop taut with a pencil tip, turn the pencil around the nails the way we do drawing a circle. After a complete turn, we get an ellipse.

Fig. C

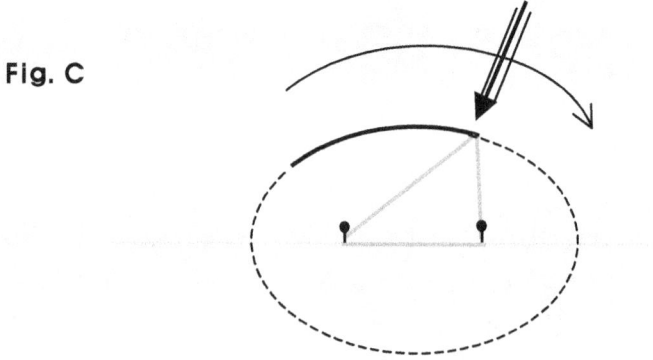

So we can see that if a point is in the ellipse, the sum of the two distances from the point to the foci is constant, that is, the same, because the length of the loop is constant.

How then, can we explain or describe a particular ellipse?

Normally, we put an ellipse in the *x-y* plain, and the ellipse is perpendicular, so one of the two main axes is perpendicular to the *x*-axis. And we can describe such an ellipse specifying the information below:

The <u>center</u> of the ellipse and the lengths of the <u>two main axes</u>, along with the <u>orientation</u>, which indicates if the ellipse is horizontal or vertical

If the major axis is horizontal, thus, parallel to the *x*-axis, the ellipse is horizontal, and if the major axis is vertical, thus, perpendicular to the *x*-axis, the ellipse is vertical.

(Note that though, it is the case in this book, and <u>it may not be the case in other books</u>. That is to say that in other books, ellipses may not be classified the way above, and thus, may not be said to be horizontal or vertical.)

Anyway, for instance, if an ellipse is <u>horizontal</u>, and is <u>centered at a point (5, 2)</u>, and its <u>main axes are 6 and 4</u>, the major axis is parallel to the *x*-axis, and the ellipse is as below:

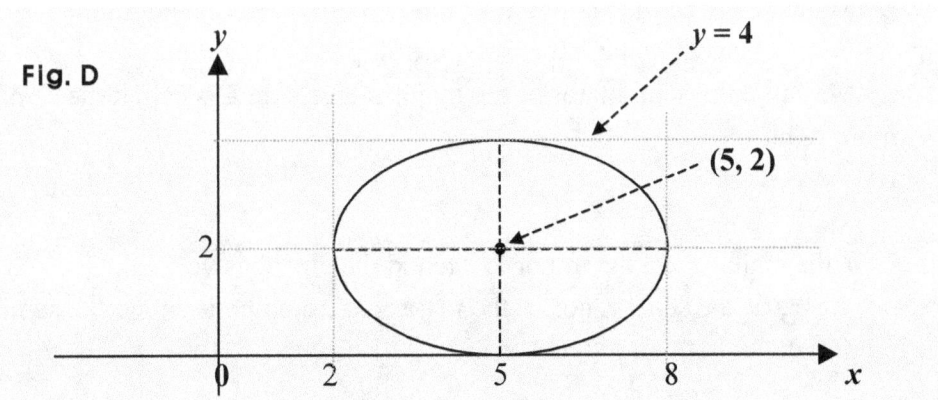

In math though, making an ellipse, we define it. So defining an ellipse particular, we make the particular ellipse.   How then, can we define a particular ellipse?

As in the cases of other conics as a circle, we can define an ellipse using an equation that indicates the ellipse. So we can define the particular ellipse producing the equation of it.

For instance, assuming E is the ellipse described above, we can define E the way below:

$$\frac{(x-5)^2}{3^2} + \frac{(y-2)^2}{2^2} = 1.$$

And the equation above is in the standard form, and is a standard equation of an ellipse.

Putting the standard equation above in the general form, we just expand (or simplify) it. Then, we get: $4x^2 + 9y^2 - 40x - 36y + 100 = 0$, which is a general equation of an ellipse.

And in general, putting an ellipse in the standard form, we can put it the way below:

$$\frac{(x-u)^2}{a^2} + \frac{(y-v)^2}{b^2} = 1, \text{ where } a \neq b, \text{ and } a \text{ and } b > 0.$$

What do we mean by though, a and b?

First, if $a > b$, $2a$ is the length of the major axis. And if we just call the length the major, the major is $2a$. Then, $2b$ is the minor, and the ellipse is horizontal.
And in that case, we call a the semi major or the major radius, and b is called the semi minor or the minor radius.

And next, if $a < b$, the major is $2b$, the minor is $2a$, and the ellipse is vertical.
And in that case, we call b the semi major or the major radius, and a is called the semi minor or the minor radius.

So just describing a particular ellipse, we can describe it specifying the main axes (or the two semis or the two radii).
If however, we define a particular ellipse placed in the x-y plane, we want to specify the center and the orientation, together with the two main axes (or the semis or the radii).

What then, is the center?

If an ellipse is: $\dfrac{(x-u)^2}{a^2} + \dfrac{(y-v)^2}{b^2} = 1$, the center of the ellipse is a point (u, v).

Note that though, the <u>standard equation</u> above can indicate an ellipse <u>perpendicular only</u>. Is there any equation then, that can indicate an ellipse <u>perpendicular or not</u>?

Putting an ellipse in such an equation, we can put it the way below:

$ax^2 + by^2 + cxy + ux + vy + w = 0$, where <u>$4ab > c^2$</u>.

Note that a and b <u>can be</u> the same, and that <u>$4ab > c^2$</u> can mean this, too: <u>$ab > 0$</u>, which is saying that a and b have the same sign. And of course, a, b, c, u, v, and w are constant. Note <u>however, for some values of the constants, the equation does not indicate an ellipse, even if $4ab > c^2$</u>.

What then, about an equation that is in the <u>general form</u> and indicates an ellipse <u>perpendicular only</u>?

Putting an ellipse in such an equation, we can put it the way below:

$ax^2 + by^2 + ux + vy + w = 0$, where <u>$a \neq b$, and $ab > 0$</u>.

So the equation above indicates an ellipse with a main axis perpendicular to the x-axis. Notice that it does not have the xy-term, that is, cxy in the equation shown earlier. Note <u>however, for some values of the constants, the equation does not indicate an ellipse, even if $a \neq b$, and $ab > 0$</u>.

And we can give a name to an equation, too. For instance, assuming E is the equation of an ellipse perpendicular above, we can put E the way as follows:

$E(x, y) = ax^2 + by^2 + ux + vy + w = 0$, where $a \neq b$, and $ab > 0$.

What then, about the standard equation?

Assuming S is the equation $\dfrac{(x-u)^2}{a^2} + \dfrac{(y-v)^2}{b^2} = 1$, we can put S the way as follows:

$$S(x, y) = \frac{(x-u)^2}{a^2} + \frac{(y-v)^2}{b^2} - 1 = 0,\text{ where } a \neq b,\text{ and } a \text{ and } b > 0.$$

What if $a = b$ though, in the equation above?

Then, the equation indicates not an ellipse but a circle.   What circle then, is it?

It is the circle where the radius is a (or b), and the center is at (u, v).

Setting: $b = a$, we get: $\dfrac{(x-u)^2}{a^2} + \dfrac{(y-v)^2}{b^2} - 1 = 0 \Rightarrow \dfrac{(x-u)^2}{a^2} + \dfrac{(y-v)^2}{a^2} = 1$

$\Rightarrow (x-u)^2 + (y-v)^2 = a^2$, which indicates a circle of radius a centered at (u, v).

If not sure of an equation of a circle, refer to **CONICS 3**.

And we know it is <u>not</u> the case the equation of the circle $(x-u)^2 + (y-v)^2 = a^2$ is defined for x real and y real. ('x real' means every real number can be x.)

It is in fact, defined for $-a + u \leq x \leq a + u$, and $-a + v \leq y \leq a + v$.

That is, the domain is: $-a + u \leq x \leq a + u$, and the range is: $-a + v \leq y \leq a + v$.

So we can notice that it is <u>not</u> the case either, the standard equation of an ellipse above is defined for x real and y real.   What then, are the domain and the range?

The domain is: $-a + u \leq x \leq a + u$, and the range is: $-b + v \leq y \leq b + v$.

So putting more specifically the standard equation of an ellipse, we can put it this way:

$\dfrac{(x-u)^2}{a^2} + \dfrac{(y-v)^2}{b^2} = 1$, where $a \neq b$, a & $b > 0$, $-a + u \leq x \leq a + u$, and $-b + v \leq y \leq b + v$.

Next, we know an ellipse can be very close to a circle, and can be very flat. So is there any way we can specify how round or flat an ellipse is?

An ellipse has a special number called an <u>eccentricity</u>, which is a ratio of a focal distance to a major radius, that is, <u>the focal distance over the major radius</u>.

So of an ellipse, the eccentricity specifies the degree, to which the ellipse is out of round. Thus, it can tell us how flat or round an ellipse is.

An eccentricity is denoted by e, and we have: **0 < e < 1** for an ellipse.
And if the eccentricity e is 0, it is a circle.

So the bigger the eccentricity, the flatter the ellipse.

If therefore, e is close to 1, the ellipse looks like a bar.

And an ellipse has a physical property as follows.

If for instance, a light beam comes from one focus, and reflects off the ellipse, it goes to the other focus. So all the light beams coming from one focus and reflecting off the ellipse gather at the other focus.

So we can use such a property constructing a reflecting telescope.

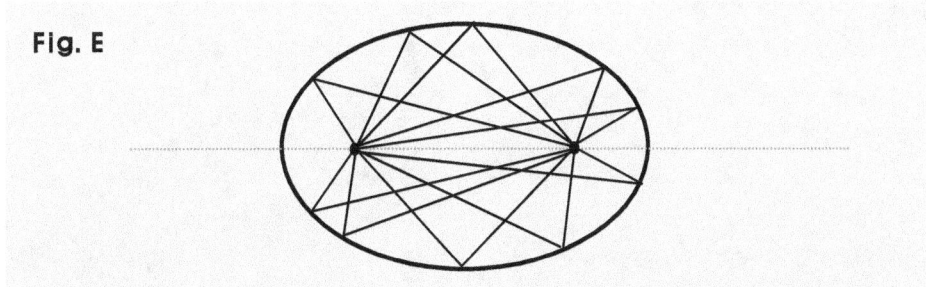

Fig. E

How then, can we get the foci and the eccentricity?

Suppose an ellipse is: $\dfrac{(x-u)^2}{a^2} + \dfrac{(y-v)^2}{b^2} = 1$.

Then, if $a > b > 0$, the foci are: $(u - c, v)$ and $(u + c, v)$, where c is the focal distance, and we have: $c^2 = a^2 - b^2$.

Then, the eccentricity $e = c/a$.   And the ellipse is horizontal.

If however, $b > a > 0$, the foci are: $(u, v + c)$ and $(u, v - c)$, where c is the focal distance, and we have: $c^2 = b^2 - a^2$.

Then, the eccentricity $e = c/b$.   And the ellipse is vertical.

Examples 1 in Standard Forms

Label each ellipse below, and then find the equation of each of the ellipses.

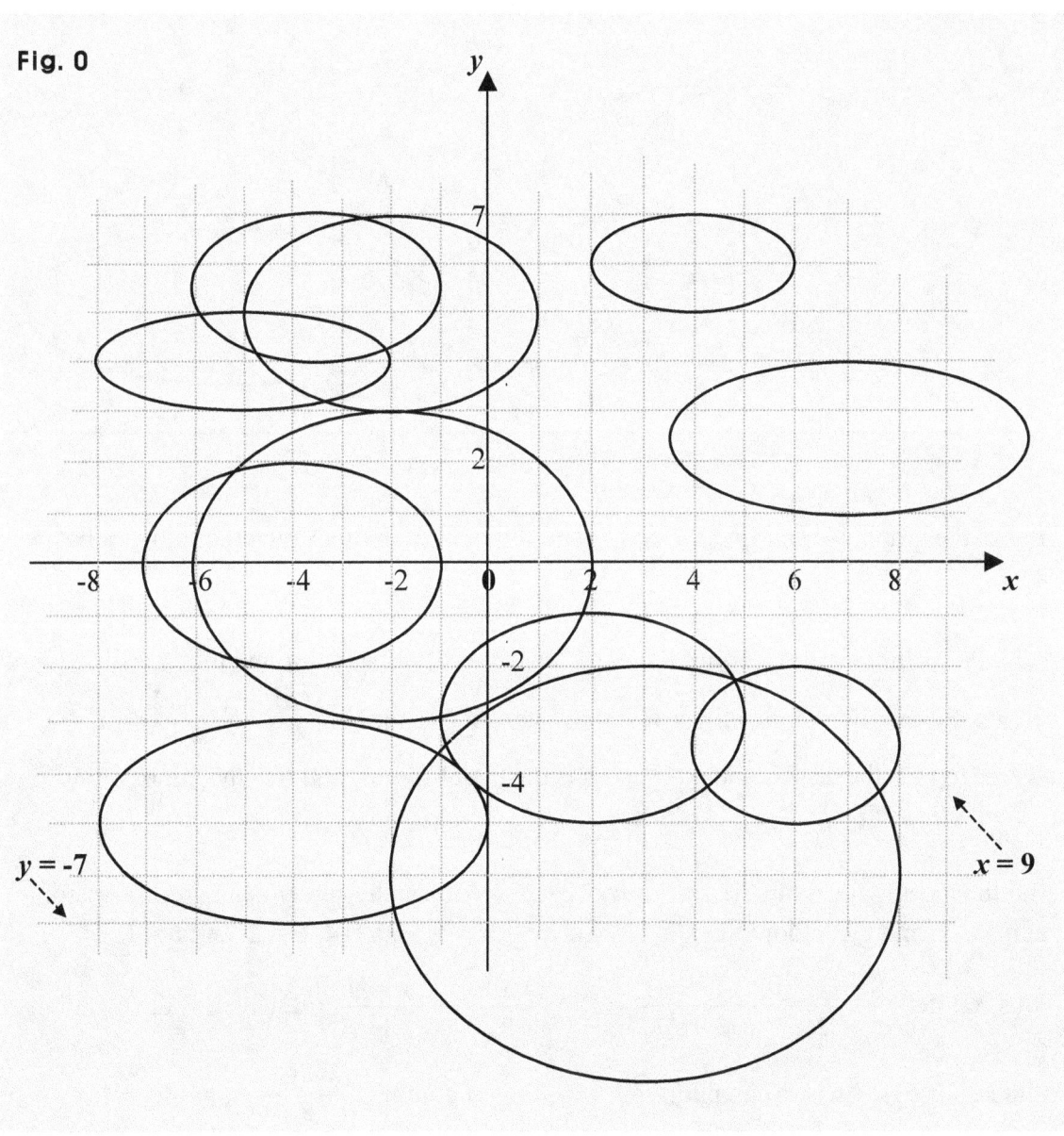

Suggestions or Solutions
To the Problems in the Examples

The ellipses are all horizontal, and beginning with the ellipses below, we can see first, the ellipse A is centered at (-6, 4), the major radius is 2, and the minor radius is 1.

Fig. 1

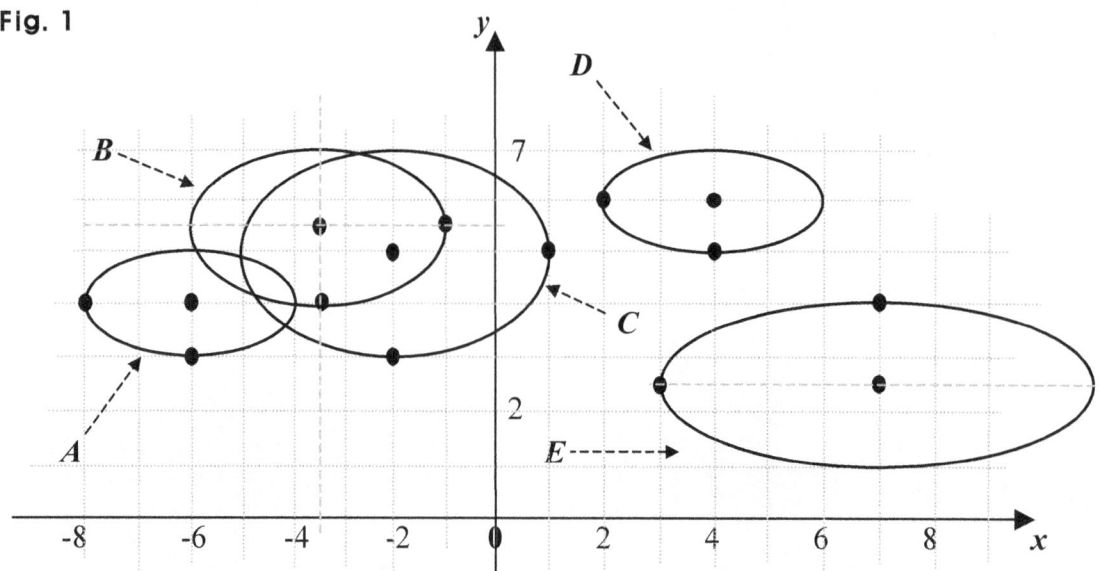

And in the graph, we can find the center and the radii each of all the other ellipses has.

Next, if an ellipse is **<u>horizontal</u>**, knowing the center and the major and minor radii, we can get the equation of the ellipse using the standard form: $\dfrac{(x-u)^2}{a^2}+\dfrac{(y-v)^2}{b^2}=1,$ where (u, v) is the center, and **<u>$a > b$</u>**, so a is the major radius, and b is the minor radius.

And thus, beginning with the ellipse A, we can see it has the center at (-6, 4), the major radius is 2, and the minor radius is 1. So we get: $u = -6$, $v = 4$, $a = 2$, and $b = 1$.

Thus, we get: $\dfrac{\{x-(-6)\}^2}{2^2}+\dfrac{(y-4)^2}{1^2}=1 \Rightarrow \dfrac{(x+6)^2}{2^2}+\dfrac{(y-4)^2}{1^2}=1,$ which is the equation

of the ellipse A. And we can put it this way, too, of course: $\dfrac{(x+6)^2}{4}+(y-4)^2=1.$

And this way, also: $(x + 6)^2 + 4(y - 4)^2 = 4.$

Next, looking at the ellipse **B**, we can see it has the center at $\left(-\frac{7}{2},\frac{11}{2}\right)$, the major radius is $\frac{5}{2}$, and the minor radius is $\frac{3}{2}$.   So we get: $u=-\frac{7}{2}, v=\frac{11}{2}, a=\frac{5}{2}$, and $b=\frac{3}{2}$.

Thus, we get: $\dfrac{\{x-(-\frac{7}{2})\}^2}{(\frac{5}{2})^2}+\dfrac{(y-\frac{11}{2})^2}{(\frac{3}{2})^2}=1 \Rightarrow \dfrac{(x+\frac{7}{2})^2}{(\frac{5}{2})^2}+\dfrac{(y-\frac{11}{2})^2}{(\frac{3}{2})^2}=1,$ which is the

equation of the ellipse **B**, and can be put this way, too: $\dfrac{(x+\frac{7}{2})^2}{\frac{25}{4}}+\dfrac{(y-\frac{11}{2})^2}{\frac{9}{4}}=1.$

And this way, also: $\dfrac{4(x+\frac{7}{2})^2}{25}+\dfrac{4(y-\frac{11}{2})^2}{9}=1$ or $\dfrac{(x+\frac{7}{2})^2}{25}+\dfrac{(y-\frac{11}{2})^2}{9}=\dfrac{1}{4}.$

Looking at next, the ellipse **C**, we can see the center is (-2, 5), the major radius is 3, and the minor radius is 2.   So we get: $u = -2, v = 5, a = 3$, and $b = 2$.

Thus, we get: $\dfrac{\{x-(-2)\}^2}{3^2}+\dfrac{(y-5)^2}{2^2}=1 \Rightarrow \dfrac{(x+2)^2}{3^2}+\dfrac{(y-5)^2}{2^2}=1,$ which is the equation

of the ellipse **C**, and can be put this way, too: $\dfrac{(x+2)^2}{9}+\dfrac{(y-5)^2}{4}=1.$

And this way, also: $4(x + 2)^2 + 9(y - 5)^2 = 36$.

Next, examining the ellipse **D**, we can see the center is (4, 6), the major radius is 2, and the minor radius is 1.   So we get: $u = 4, v = 6, a = 2$, and $b = 1$.

Thus, we get: $\dfrac{(x-4)^2}{2^2}+\dfrac{(y-6)^2}{1^2}=1,$ which is the equation of the ellipse **D**, and can be

put this way, too: $\dfrac{(x-4)^2}{4}+(y-6)^2=1.$ And this way, also: $(x-4)^2+4(y-6)^2=4.$

And next, looking at the ellipse **E**, we can see the center is $\left(7,\frac{5}{2}\right)$, the major radius is $\frac{7}{2}$, and the minor radius is $\frac{3}{2}$. So we get: $u = 7, v=\frac{5}{2}, a=\frac{7}{2}$, and $b=\frac{3}{2}.$

Thus, we get: $\dfrac{(x-7)^2}{(\frac{7}{2})^2}+\dfrac{(y-\frac{5}{2})^2}{(\frac{3}{2})^2}=1$, which is the equation of **E**, and can be put this

way, too: $\dfrac{(x-7)^2}{\frac{49}{4}}+\dfrac{(y-\frac{5}{2})^2}{\frac{9}{4}}=1$, $\dfrac{4(x-7)^2}{49}+\dfrac{4(y-\frac{5}{2})^2}{9}=1$ or $\dfrac{(x-7)^2}{49}+\dfrac{(y-\frac{5}{2})^2}{9}=\dfrac{1}{4}$.

Next, moving on to the ellipses below, we can see that the ellipse **F** is centered at (-4, 0), the major radius is 3, and the minor radius is 2.

Fig. 2

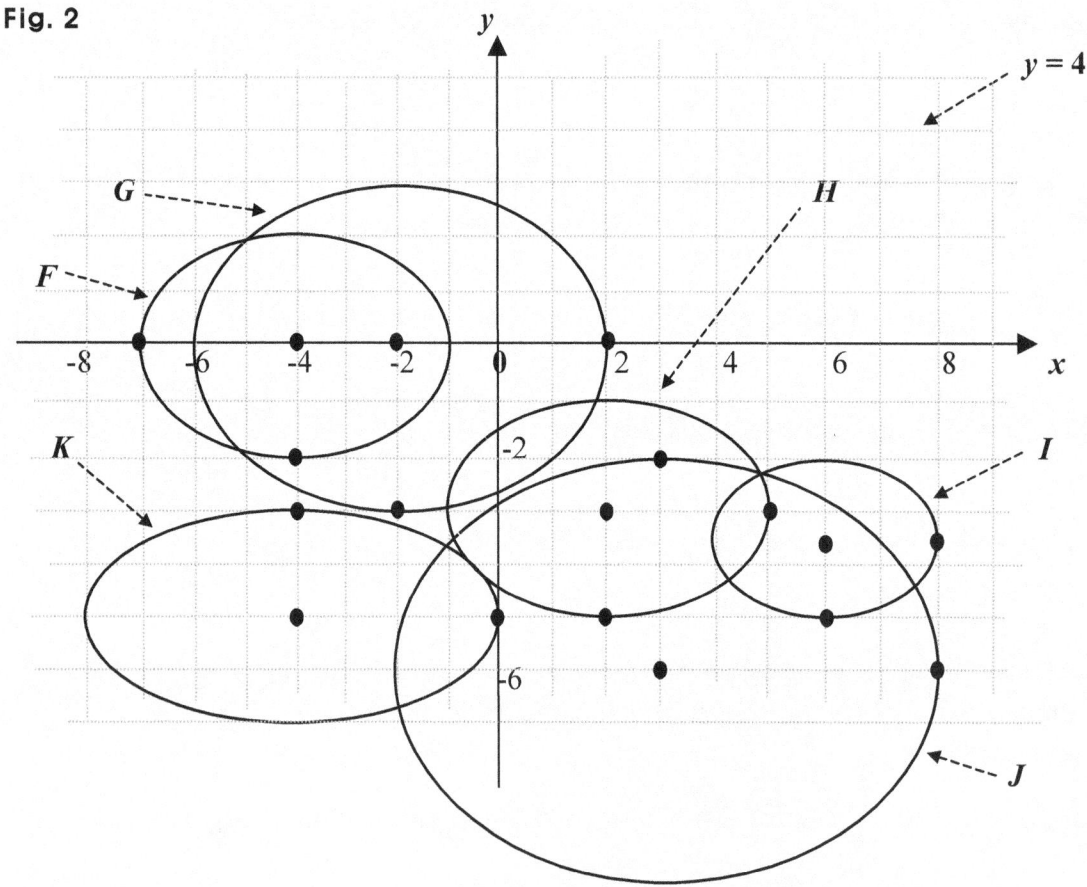

And if an ellipse is **<u>horizontal</u>**, knowing the center and the major and minor radii, we

can get the equation of the ellipse using the standard form: $\dfrac{(x-u)^2}{a^2}+\dfrac{(y-v)^2}{b^2}=1$,

where **(u, v)** is the center, and **<u>a > b</u>**, so **a** is the major radius, and **b** is the minor radius.

And thus, beginning with the ellipse **F**, we can see the center is (-4, 0), the major radius is 3, and the minor radius is 2. So we get: $u = -4$, $v = 0$, $a = 3$, and $b = 2$.

Thus, we get: $\dfrac{\{x-(-4)\}^2}{3^2}+\dfrac{(y-0)^2}{2^2}=1 \Rightarrow \dfrac{(x+4)^2}{3^2}+\dfrac{y^2}{2^2}=1$, which is the equation of **F**,

and can be put the way as follows, too: $\dfrac{(x+4)^2}{9}+\dfrac{y^2}{4}=1$, or $4(x+4)^2 + 9y^2 = 36$.

Next, looking at the ellipse **G**, we can see the center is (-2, 0), the major radius is 4, and the minor radius is 3. So we get: $u = -2$, $v = 0$, $a = 4$, and $b = 3$.

Thus, we get: $\dfrac{\{x-(-2)\}^2}{4^2}+\dfrac{(y-0)^2}{3^2}=1 \Rightarrow \dfrac{(x+2)^2}{4^2}+\dfrac{y^2}{3^2}=1$, which is the equation of **G**,

and can be put this way, too: $\dfrac{(x+2)^2}{16}+\dfrac{y^2}{9}=1$, or this way: $9(x+2)^2 + 16(y-6)^2 = 144$.

Next, examining the ellipse **H**, we can see the center is (2, -3), the major radius is 3, and the minor radius is 2. So we get: $u = 2$, $v = -3$, $a = 3$, and $b = 2$.

Thus, we get: $\dfrac{(x-2)^2}{3^2}+\dfrac{\{y-(-3)\}^2}{2^2}=1 \Rightarrow \dfrac{(x-2)^2}{3^2}+\dfrac{(y+3)^2}{2^2}=1$, which is the equation

of **H**, and can be put this way, too: $\dfrac{(x-2)^2}{9}+\dfrac{(y+3)^2}{4}=1$ or $4(x-2)^2 + 9(y+3)^2 = 36$.

Next, examining the ellipse **I**, we can see the center is $(6, \frac{7}{2})$, the major radius is 2, and the minor radius is $\frac{3}{2}$. So we get: $u = 6$, $v = \frac{7}{2}$, $a = 2$, and $b = \frac{3}{2}$.

Thus, we get: $\dfrac{(x-6)^2}{2^2}+\dfrac{(y-\frac{7}{2})^2}{(\frac{3}{2})^2}=1$, which is the equation of **E**, and can be put this

way, too: $\dfrac{(x-6)^2}{4}+\dfrac{(y-\frac{7}{2})^2}{\frac{9}{4}}=1$, $\dfrac{(x-6)^2}{4}+\dfrac{4(y-\frac{7}{2})^2}{9}=1$, or $9(x-6)^2 + 16(y-\frac{7}{2})^2 = 36$.

20

Fig. 3

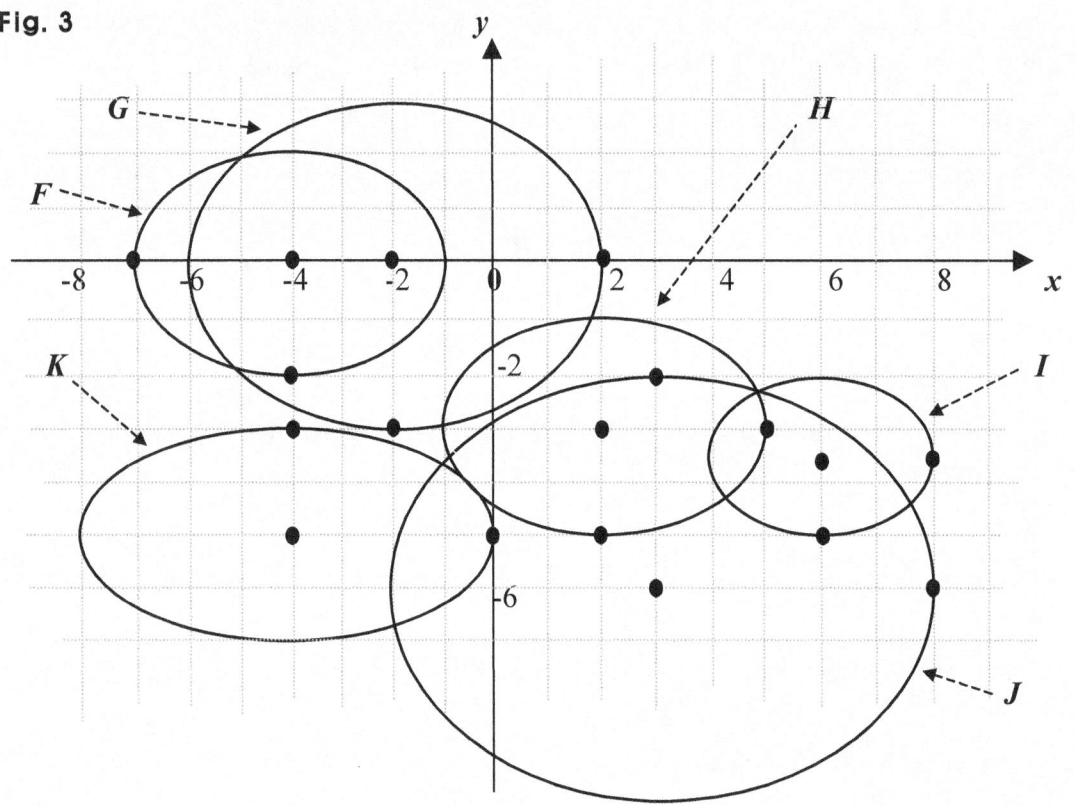

Next, examining the ellipse **J**, we can see the center is (3, -6), the major radius is 5, and the minor radius is 4.   So we get: **u = 3**, **v = -6**, **a = 5**, and **b = 4**.

Thus, we get: $\dfrac{(x-3)^2}{5^2}+\dfrac{\{y-(-6)\}^2}{4^2}=1 \Rightarrow \dfrac{(x-3)^2}{5^2}+\dfrac{(y+6)^2}{4^2}=1$, which is the ellipse **J**,

and can be put this way, too: $\dfrac{(x-3)^2}{25}+\dfrac{(y+6)^2}{16}=1$ or $16(x-3)^2+25(y+3)^2=400$.

Next, examining the ellipse **K**, we can see the center is (-4, -5), the major radius is 4, and the minor radius is 2.   So we get: **u = -4**, **v = -5**, **a = 4**, and **b = 2**.

Thus, we get: $\dfrac{\{x-(-4)\}^2}{4^2}+\dfrac{\{y-(-5)\}^2}{2^2}=1 \Rightarrow \dfrac{(x+4)^2}{4^2}+\dfrac{(y+5)^2}{2^2}=1$, which is the ellipse

K, and can be put this way, too: $\dfrac{(x+4)^2}{16}+\dfrac{(y+5)^2}{4}=1$ or $4(x+4)^2+16(y+5)^2=64$.

Examples 2 in Standard Forms

Label each ellipse below, and then find the equation of each of the ellipses.

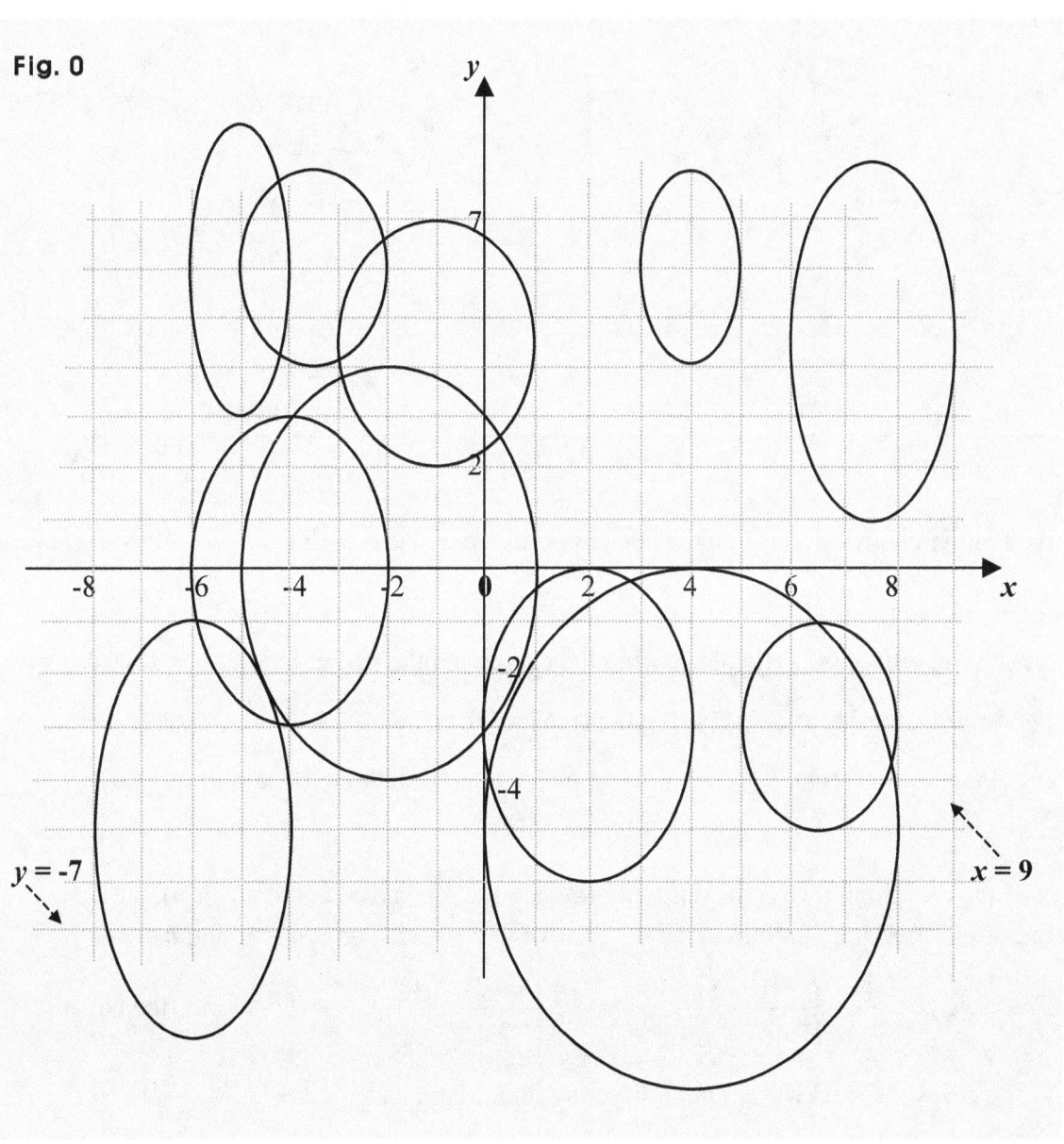

Fig. 0

Suggestions or Solutions
To the Problems in the Examples

The ellipses are all vertical, and beginning with the ellipses below, we can see first, the ellipse *A* is centered at (-5, 6), the major radius is 3, and the minor radius is 1.

Fig. 1

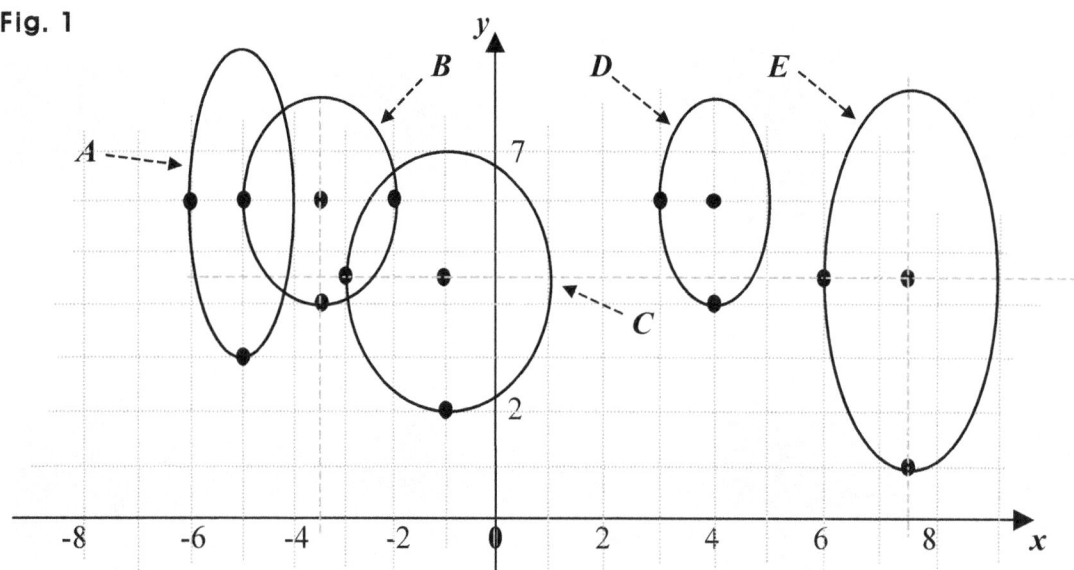

And in the graph, we can find the center and the radii each of all the other ellipses has.

Next, if an ellipse is **vertical**, knowing the center and the major and minor radii, we can get the equation of the ellipse using the standard form: $\dfrac{(x-u)^2}{a^2}+\dfrac{(y-v)^2}{b^2}=1,$ where *(u, v)* is the center, and **_b > a_**, so *b* is the major radius, and *a* is the minor radius.

And thus, beginning with the ellipse *A*, we can see it has the center at (-5, 6), the major radius is 3, and the minor radius is 1.   So we get: ***u* = -5, *v* = 6, *a* = 1**, and ***b* = 3**.

Thus, we get: $\dfrac{\{x-(-5)\}^2}{1^2}+\dfrac{(y-6)^2}{3^2}=1 \Rightarrow \dfrac{(x+5)^2}{1^2}+\dfrac{(y-6)^2}{3^2}=1,$ which is the equation of the ellipse *A*. And we can put it this way, too, of course: $(x+5)^2+\dfrac{(y-6)^2}{9}+=1.$ And this way, also: $9(x+5)^2+(y-6)^2=9.$

Next, looking at the ellipse **B**, we can see it has the center at $(-\frac{7}{2}, 6)$, the major radius is 2, and the minor radius is $\frac{3}{2}$.   So we get: $u = -\frac{7}{2}$, $v = 6$, $a = \frac{3}{2}$, and $b = 2$.

Thus, we get: $\dfrac{\{x-(-\frac{7}{2})\}^2}{(\frac{3}{2})^2} + \dfrac{(y-6)^2}{2^2} = 1 \Rightarrow \dfrac{(x+\frac{7}{2})^2}{(\frac{3}{2})^2} + \dfrac{(y-6)^2}{2^2} = 1$, which is the equation

of the ellipse **B**, and can be put this way, too: $\dfrac{(x+\frac{7}{2})^2}{\frac{9}{4}} + \dfrac{(y-6)^2}{4} = 1$.

And this way, also: $\dfrac{4(x+\frac{7}{2})^2}{9} + \dfrac{(y-6)^2}{4} = 1$, or $16(x+\frac{7}{2})^2 + 9(y-6)^2 = 36$.

Looking at next, the ellipse **C**, we can see the center is $(-1, \frac{9}{2})$, the major radius is $\frac{5}{2}$, and the minor radius is 2. So we get: $u = -1$, $v = \frac{9}{2}$, $a = 2$, and $b = \frac{5}{2}$.

Thus, we get: $\dfrac{(x+1)^2}{2^2} + \dfrac{(y-\frac{9}{2})^2}{(\frac{5}{2})^2} = 1$, which is the equation of **C**, and can be put this way,

too: $\dfrac{(x+1)^2}{4} + \dfrac{(y-\frac{9}{2})^2}{\frac{25}{4}} = 1$, $\dfrac{(x+1)^2}{4} + \dfrac{4(y-\frac{9}{2})^2}{25} = 1$, or $25(x+1)^2 + 16(y-\frac{9}{2})^2 = 100$.

Next, examining the ellipse **D**, we can see the center is (4, 6), the major radius is 2, and the minor radius is 1.   So we get: $u = 4$, $v = 6$, $a = 1$, and $b = 2$.

Thus, we get: $\dfrac{(x-4)^2}{1^2} + \dfrac{(y-6)^2}{2^2} = 1$, which is the equation of the ellipse **D**, and can be

put this way, too: $(x-4)^2 + \dfrac{(y-6)^2}{4} = 1$.   And this way, also: $4(x-4)^2 + (y-6)^2 = 4$.

And next, looking at the ellipse **E**, we can see the center is $(\frac{15}{2}, \frac{9}{2})$, the major radius is $\frac{7}{2}$, and the minor radius is $\frac{3}{2}$.   So we get: $u = \frac{15}{2}$, $v = \frac{9}{2}$, $a = \frac{3}{2}$, and $b = \frac{7}{2}$.

Thus, we get: $\dfrac{(x-\frac{15}{2})^2}{(\frac{3}{2})^2}+\dfrac{(y-\frac{9}{2})^2}{(\frac{7}{2})^2}=1$, which is the equation of **E**, and can be put this

way, too: $\dfrac{(x-\frac{15}{2})^2}{\frac{9}{4}}+\dfrac{(y-\frac{9}{2})^2}{\frac{49}{4}}=1$, $\dfrac{4(x-\frac{15}{2})^2}{9}+\dfrac{4(y-\frac{9}{2})^2}{49}=1$ or $\dfrac{(x-\frac{15}{2})^2}{9}+\dfrac{(y-\frac{9}{2})^2}{49}=\dfrac{1}{4}$.

Next, moving on to the ellipses below, we can see that the ellipse **F** is centered at (-4, 0), and a point at (-7, 4) or (-3, 4).

Fig. 2

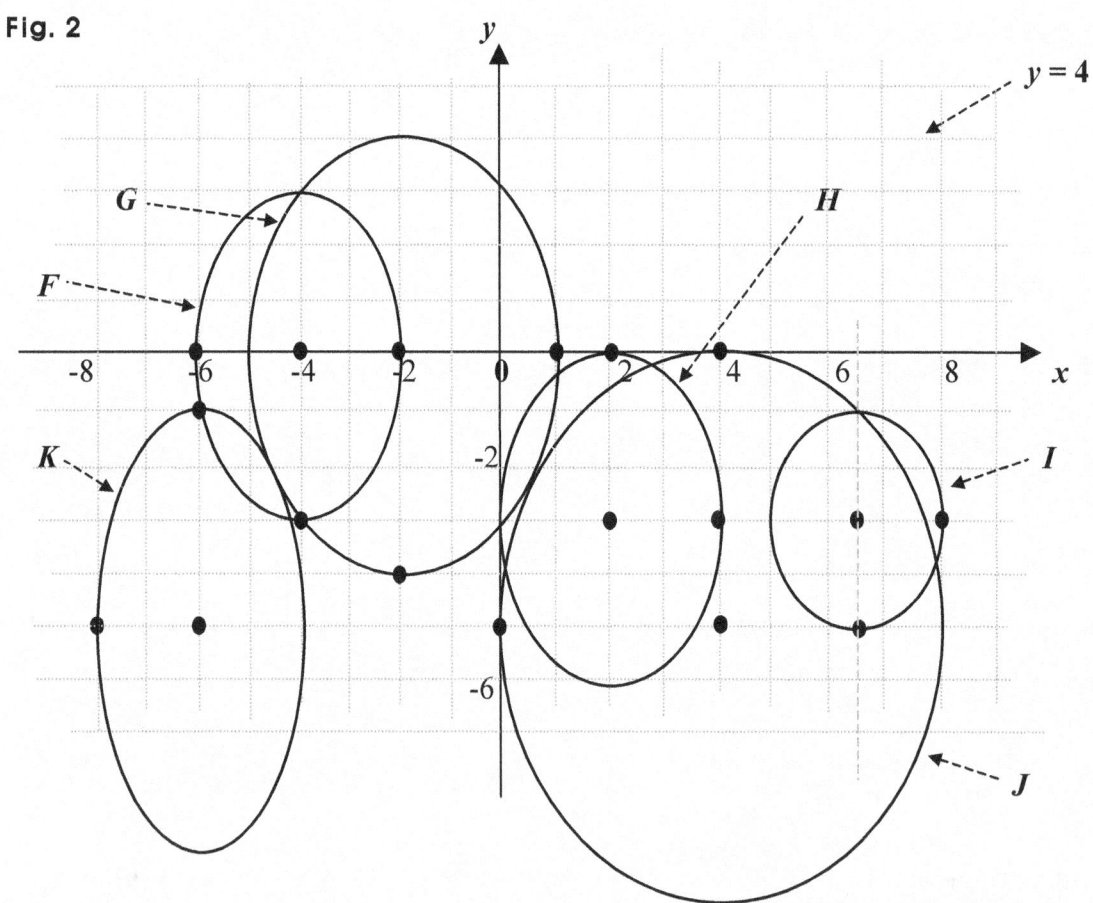

And if an ellipse is **horizontal**, knowing the center and the major and minor radii, we

can get the equation of the ellipse using the standard form: $\dfrac{(x-u)^2}{a^2}+\dfrac{(y-v)^2}{b^2}=1$,

where **(u, v)** is the center, and **_a > b_**, so **a** is the major radius, and **b** is the minor radius.

And thus, beginning with the ellipse **F**, we can see the center is (-4, 0), the major radius is 3, and the minor radius is 2.   So we get: **u = -4, v = 0, a = 2,** and **b = 3.**

Thus, we get: $\dfrac{\{x-(-4)\}^2}{2^2}+\dfrac{(y-0)^2}{3^2}=1 \Rightarrow \dfrac{(x+4)^2}{2^2}+\dfrac{y^2}{3^2}=1,$ which is the equation of **F**,

and can be put this way below, too: $\dfrac{(x+4)^2}{4}+\dfrac{y^2}{9}=1,$ or this way: $9(x+4)^2+4y^2=36.$

Next, looking at the ellipse **G**, we can see the center is (-2, 0), the major radius is 4, and the minor radius is 3.   So we get: **u = -2, v = 0, a = 3,** and **b = 4.**

Thus, we get: $\dfrac{\{x-(-2)\}^2}{3^2}+\dfrac{(y-0)^2}{4^2}=1 \Rightarrow \dfrac{(x+2)^2}{3^2}+\dfrac{y^2}{4^2}=1,$ which is the equation of **G**,

and can be put this way, too: $\dfrac{(x+2)^2}{9}+\dfrac{y^2}{16}=1,$ or this way: $16(x+2)^2+9(y-6)^2=144.$

Next, examining the ellipse **H**, we can see the center is (2, -3), the major radius is 3, and the minor radius is 2.   So we get: **u = 2, v = -3, a = 2,** and **b = 3.**

Thus, we get: $\dfrac{(x-2)^2}{2^2}+\dfrac{\{y-(-3)\}^2}{3^2}=1 \Rightarrow \dfrac{(x-2)^2}{2^2}+\dfrac{(y+3)^2}{3^2}=1,$ which is the equation

of **H**, and can be put this way, too: $\dfrac{(x-2)^2}{4}+\dfrac{(y+3)^2}{9}=1$ or $9(x-2)^2+4(y+3)^2=36.$

Next, examining the ellipse **I**, we can see the center is $(\frac{13}{2},-3),$ the major radius is 2, and the minor radius is $\frac{3}{2}.$ So we get: $u=\frac{13}{2}, v=-3, a=\frac{3}{2},$ and **b = 2.**

Thus, we get: $\dfrac{(x-\frac{13}{2})^2}{(\frac{3}{2})^2}+\dfrac{(y+3)^2}{2^2}=1,$ which is the equation of **E**, and can be put this

way, too: $\dfrac{(x-\frac{13}{2})^2}{\frac{9}{4}}+\dfrac{(y+3)^2}{4}=1, \dfrac{4(x-\frac{13}{2})^2}{9}+\dfrac{(y+3)^2}{4}=1,$ or $16(x-\frac{13}{2})^2+9(y+3)^2=36.$

Fig. 3

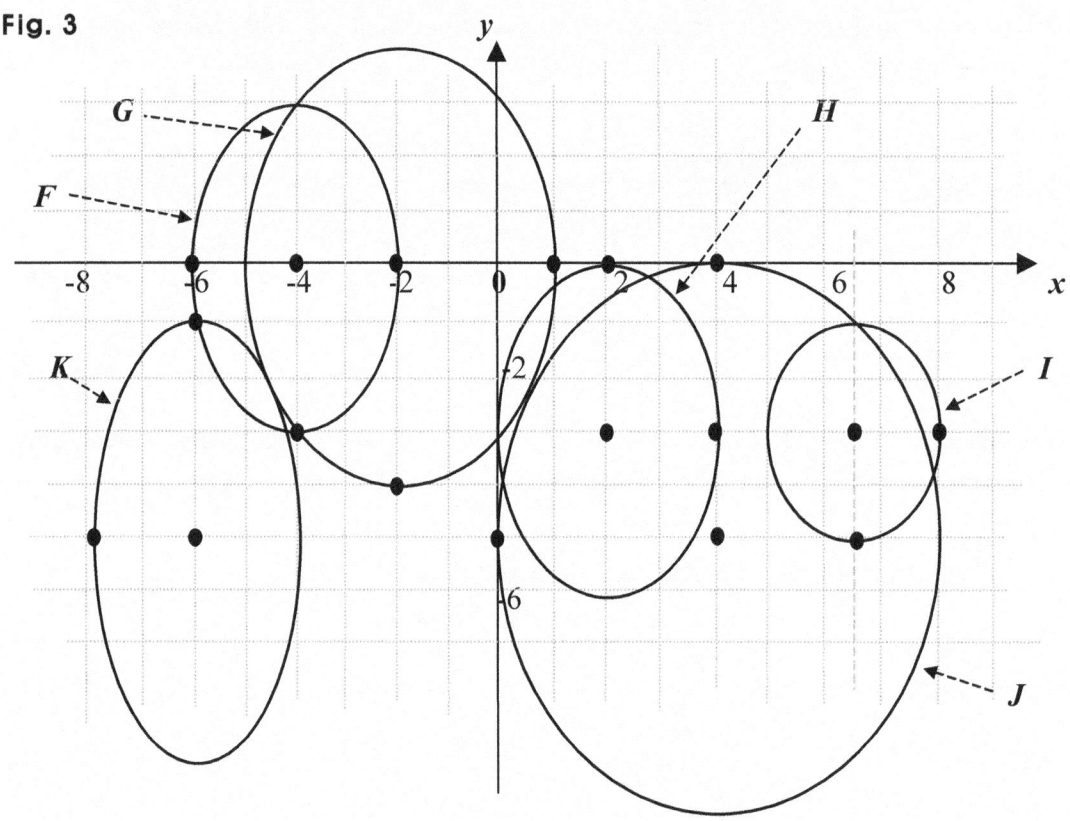

Next, examining the ellipse **J**, we can see the center is (4, -5), the major radius is 5, and the minor radius is 4.   So we get: **u = 4**, **v = -5**, **a = 4**, and **b = 5**.

Thus, we get: $\dfrac{(x-4)^2}{4^2}+\dfrac{\{y-(-5)\}^2}{5^2}=1 \Rightarrow \dfrac{(x-4)^2}{4^2}+\dfrac{(y+5)^2}{5^2}=1$, which is the ellipse **J**,

and can be put this way, too: $\dfrac{(x-4)^2}{16}+\dfrac{(y+5)^2}{25}=1$ or **25(x − 4)² + 16(y + 5)² = 400**.

Next, examining the ellipse **K**, we can see the center is (-6, -5), the major radius is 4, and the minor radius is 2.   So we get: **u = -6**, **v = -5**, **a = 2**, and **b = 4**.

Thus, we get: $\dfrac{\{x-(-6)\}^2}{2^2}+\dfrac{\{y-(-5)\}^2}{4^2}=1 \Rightarrow \dfrac{(x+6)^2}{2^2}+\dfrac{(y+5)^2}{4^2}=1$, which is the ellipse

K, and can be put this way, too: $\dfrac{(x+6)^2}{4}+\dfrac{(y+5)^2}{16}=1$ or **16(x + 6)² + 4(y + 5)² = 64**.

1. Equations for Ellipses 1

To begin with, putting an ellipse in an equation, we can get:

$$\frac{(x-5)^2}{3^2} + \frac{(y-2)^2}{2^2} = 1.$$

What ellipse then, is it?

It is an ellipse <u>horizontal</u>, the <u>center is at (5, 2)</u>, and its <u>main axes are 6 and 4</u>. So the major axis is parallel to the *x*-axis, and is 6, and the ellipse is as below:

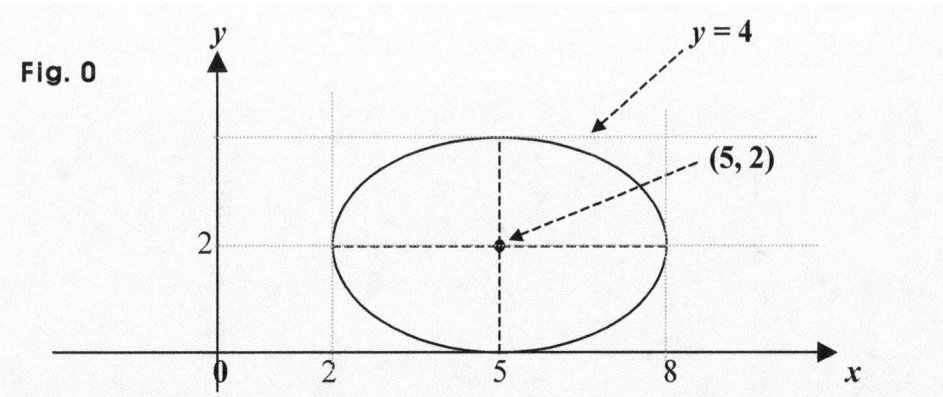

Fig. 0

What ellipse then, is the ellipse below?

$$\frac{x^2}{3^2} + \frac{y^2}{2^2} = 1.$$

The ellipse is <u>horizontal</u>, too, and its <u>main axes are 6 and 4</u>, also, but the <u>center is</u> <u>different, and is at (0, 0)</u>, that is, at the origin.

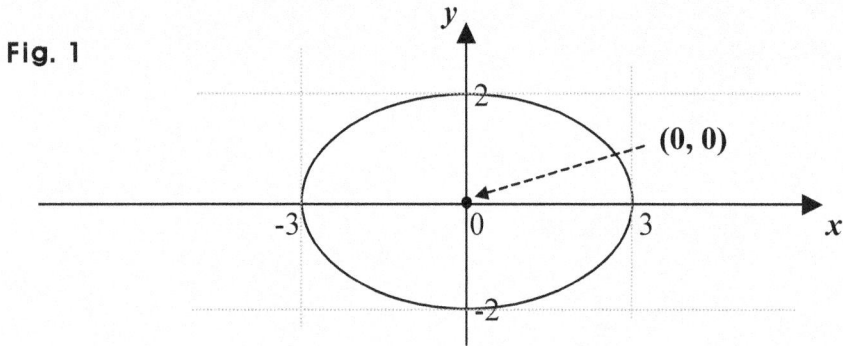

Fig. 1

What if the center is at (1, 1), and the other information is the same?

Then, the ellipse will be as follows: $\dfrac{(x-1)^2}{3^2}+\dfrac{(y-1)^2}{2^2}=1$.

And we can put the ellipse above in a graph the way below:

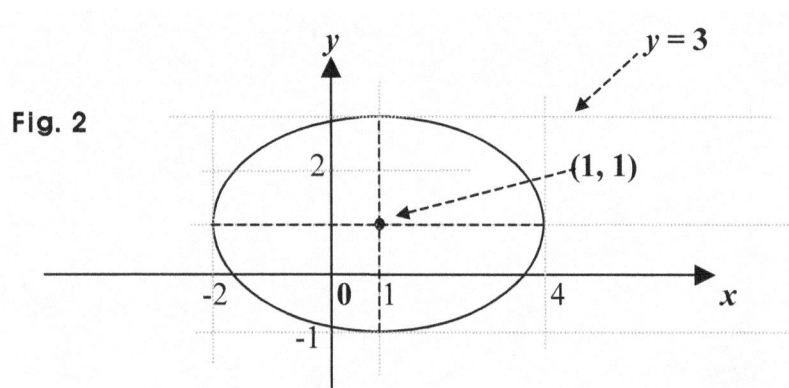

Fig. 2

How then, can we get such equations as above?

We know the fact that an ellipse is a set of points, from each of which, the sum of the two distances to two points called the foci is constant.   So?

So using the fact above, we can get the equation of an ellipse.

Thus, to begin with, suppose $c > 0$, two points $P(-c, 0)$ and $Q(c, 0)$ are the two foci of an ellipse called E, and a point $T(x, y)$ is an arbitrary point in the ellipse E, that is, a random point representing all the points in E.

Then, the center is the origin, that is, $(0, 0)$, and c is the focal distance, because the center is the midpoint between the foci. So we can put the three points P, Q, and T in the x-y plane the way as follows:

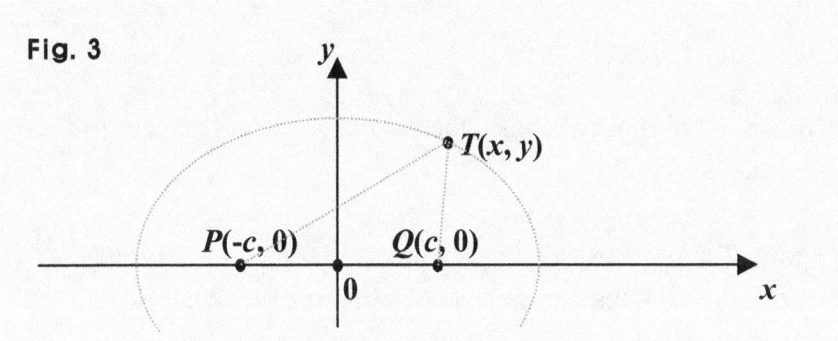

Fig. 3

Suppose next, the point T is moving along the ellipse E, and is now at $(a, 0)$.
Then, we get:

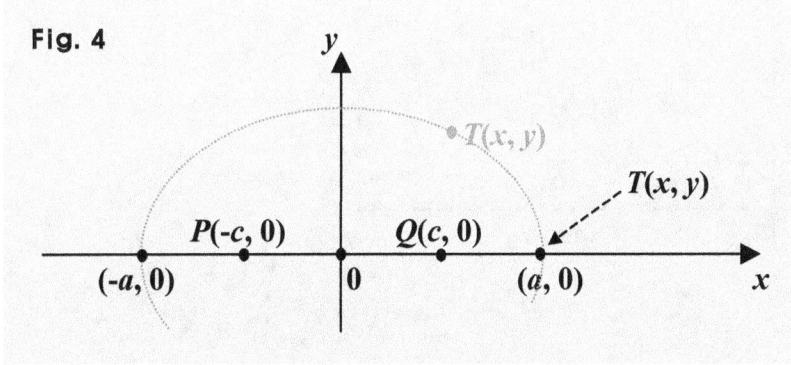

Fig. 4

Then, the sum of the two distances from $T(x, y)$ to the foci is $2a$.   How come?

Fig. 5

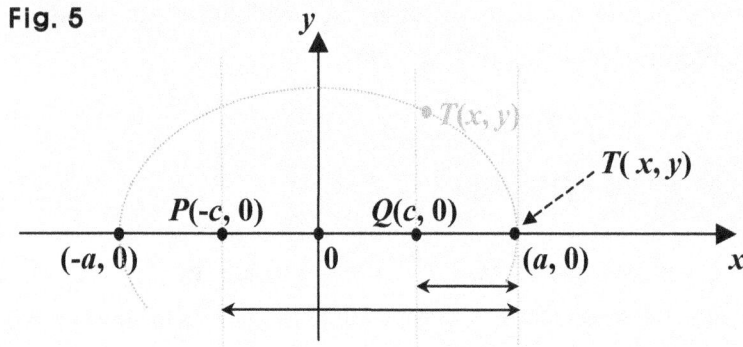

We can see that the length of **TP** is: **a – (-c) = a + c**.

And the length of **TQ** is: **a – c**.

So we get: **TP + TQ = (a + c) + (a – c) = 2a**.   So what?

We know the fact that from any point in an ellipse, the sum of the two distances to two points called the foci is constant.   So?

So no matter where the point **T** may be in the ellipse **E**, the sum of the two distances from the point **T** to the two foci is always the same, and is in this case, **2a**.

Using thus, the fact above, we can get the equation of **E**. How then, can we use the fact?

Fig. 6

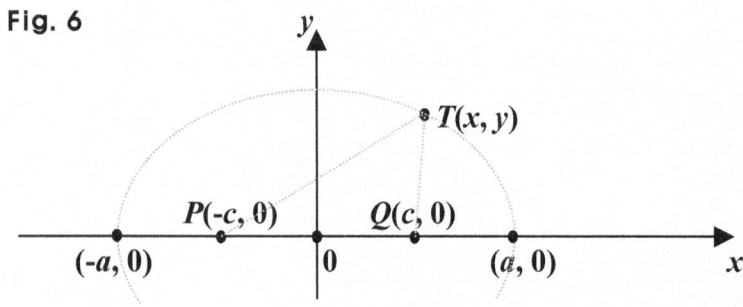

We are going to get the sum of the two distances from **T(x, y)** to the two foci, and then, set the sum equal to the sum we found above, which is **2a**, because the sum is constant, that is, the same no matter where the point **T** may be in the ellipse **E**.

How then, can we get the two distances from $T(x, y)$ to the two foci?

We can use the distance formula, often called Pythagorean Theorem.

What then, do we get?

We get an equation expressed in terms of the coordinates of the point T, that is, we get an equation in terms of x and y. And we call such an equation a connective equation.

So in this case, the connective equation is the equation that connects the two variables used as the coordinates of the arbitrary point $T(x, y)$ in the curve called the ellipse E.

How then, do we call the equation?

The equation explains every point in the ellipse E, that is, it indicates the ellipse E. So the equation is called the equation of the ellipse E.

So let's now get the equation.   How then, can we apply the distance formula?

Assuming d is the distance between two points, Δx is the difference in x-coordinates, and Δy is the difference in y-coordinates, we can put the formula the way below:

$$d^2 = (\Delta x)^2 + (\Delta y)^2.$$

So using the distance formula, and assuming $TP = p$, and $TQ = q$, we can get:

$$p^2 = \{x - (-c)\}^2 + (y - 0)^2 = (x + c)^2 + y^2, \text{ and } q^2 = (x - c)^2 + (y - 0)^2 = (x - c)^2 + y^2.$$

So we get: $TP = \sqrt{(x + c)^2 + y^2}$, and $TQ = \sqrt{(x - c)^2 + y^2}$.

Thus, we get: $TP + TQ = \sqrt{(x + c)^2 + y^2} + \sqrt{(x - c)^2 + y^2} = 2a$. Then, we get:

$$\sqrt{(x+c)^2 + y^2} = 2a - \sqrt{(x-c)^2 + y^2} \Rightarrow (x+c)^2 + y^2 = (2a - \sqrt{(x-c)^2 + y^2})^2$$

$$\Rightarrow (x+c)^2 + y^2 = 4a^2 - 4a\sqrt{(x-c)^2 + y^2} + (x-c)^2 + y^2$$

$$\Rightarrow 4a\sqrt{(x-c)^2 + y^2} = (x-c)^2 + y^2 - (x+c)^2 - y^2 + 4a^2$$

$$\Rightarrow 4a\sqrt{(x-c)^2 + y^2} = x^2 - 2cx + c^2 - x^2 - 2cx - c^2 + 4a^2 = 4a^2 - 4cx$$

$$\Rightarrow a\sqrt{(x-c)^2 + y^2} = a^2 - cx \Rightarrow a^2\{(x-c)^2 + y^2\} = (a^2 - cx)^2 = a^4 - 2a^2cx + c^2x^2$$

$$\Rightarrow a^2(x^2 - 2cx + c^2 + y^2) = a^2x^2 - 2a^2cx + a^2c^2 + a^2y^2 = a^4 - 2a^2cx + c^2x^2$$
$$\Rightarrow a^2x^2 - c^2x^2 + a^2c^2 + a^2y^2 = a^4 \Rightarrow (a^2 - c^2)x^2 + a^2y^2 = a^4 - a^2c^2 = a^2(a^2 - c^2)$$

$$\Rightarrow x^2 + \frac{a^2y^2}{a^2 - c^2} = a^2 \Rightarrow \frac{x^2}{a^2} + \frac{y^2}{a^2 - c^2} = 1. \quad \text{What then, about } a^2 - c^2?$$

It is constant, so assuming *b* is constant, we can set: $b^2 = a^2 - c^2$. And assuming *b* > 0, we can put two points **(0, b)** and **(0, -b)** the way below:

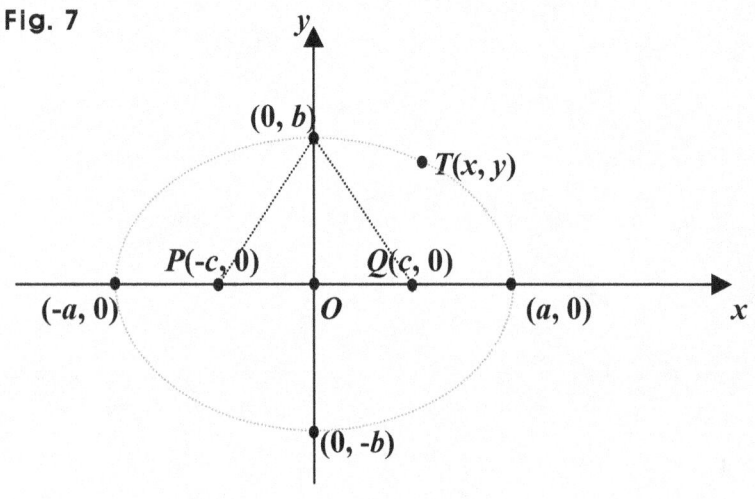

Fig. 7

How come though?

Assuming now, that the point T is at $(0, b)$, we can see a triangle isosceles, and the triangle is $\triangle TPQ$.

And we know that the sum of the two distances from T to the foci is $2a$.

That is, we have: $TP + TQ = 2a$.   So we get: $TP = TQ = a$.   So?

Assuming O is the origin, we can see that the triangle TOP is a right triangle, and its hypotenuse TP is a, and the side PO is c.

So applying the distance formula, we can get: $a^2 = b^2 + c^2 \Rightarrow b^2 = a^2 - c^2$.

And we now have: $\dfrac{x^2}{a^2} + \dfrac{y^2}{a^2 - c^2} = 1$.

So we get: $\dfrac{x^2}{a^2} + \dfrac{y^2}{b^2} = 1$, and we know: $a > b > 0$, since $a^2 = b^2 + c^2 \Rightarrow a > b > 0$.

And the equation above is expressed in terms of the coordinates of the point $T(x, y)$, so it is an equation in terms of x and y. And we call such an equation a connective equation.

So in this case, the connective equation is the equation that connects the two variables used as the coordinates of the arbitrary point $T(x, y)$ in the curve called the ellipse E.

Thus, the equation of the ellipse E is: $\dfrac{x^2}{a^2} + \dfrac{y^2}{b^2} = 1$, where $a > b > 0$.

And putting the ellipse E in the x-y plane, we get:

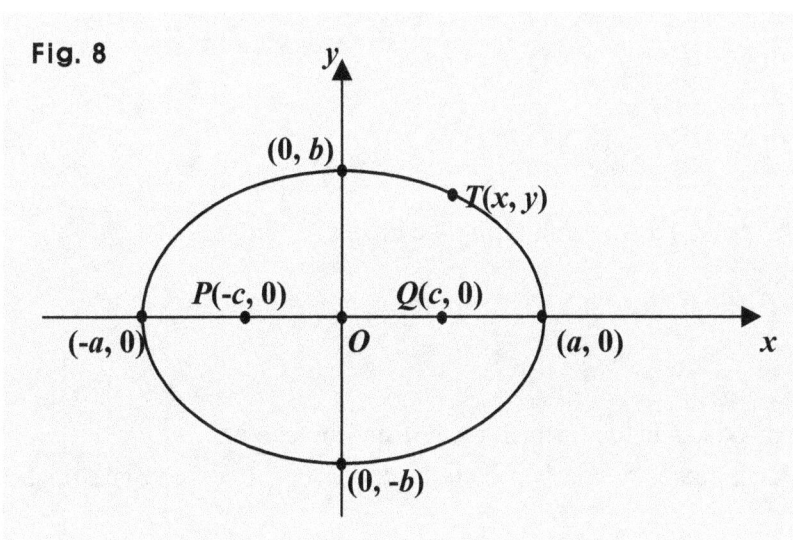

Fig. 8

So we can now say that the ellipse *E* is horizontal and centered at the origin, and that the major axis is **2a**, and the minor axis is **2b**.

That is, the semi major axis is *a*, and the semi minor is *b*. In other words, the major radius is *a*, and is parallel to the *x*-axis, and the minor radius is *b*.

And putting the ellipse *E* in its equation, we get: $\dfrac{x^2}{a^2} + \dfrac{y^2}{b^2} = 1$, where $a > b > 0$.

What then, are the domain and the range?

The domain is $-a \le x \le a$, that is, $|x| \le a$, and the range is: $-b \le y \le b$ or $|y| \le b$.

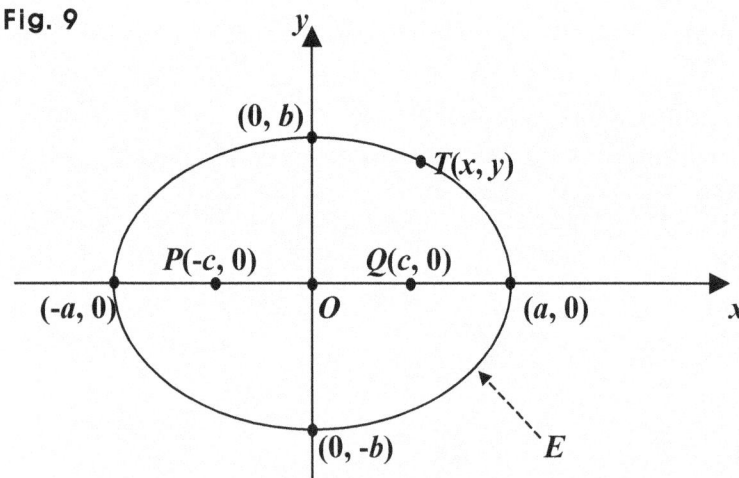

Fig. 9

Suppose now again, that the point *T* is moving along the ellipse *E* above.

Then, when will the point *T* be the farthest away from the center?

When the point *T* is in the *x*-axis, *T* is the farthest away from the center.
And we know that the major radius is *a*, and is parallel to the *x*-axis, because the ellipse *E* is horizontal.

So the major radius is a part of the *x*-axis, and *a* is the distance from *T* to the center when *T* is the farthest away from the center.

And the ellipse has two points where *T* gets located when *T* is the farthest away from the center. And the two are: **(-a, 0)** and **(a, 0)**, which are called antipodal points.

Antipodal points are two points, and are in a line passing through the center located between the two points. So they are a pair of points facing each other over the center.

And in the case of an ellipse or a circle, antipodal points are the same distance away from the center. So in that case, the center is the midpoint between the antipodal points.

So the major axis is the line segment connecting the two antipodal points where *T* gets positioned when *T* is the farthest away from the center, and the two are: **(-a, 0)** and **(a, 0)**.

Thus, the two antipodal points **(-a, 0)** and **(a, 0)** are the endpoints of the major axis, and the length of it is **2a**.

What then, about the case where the point *T* is the closest to the center?

When the point *T* is in the *y*-axis, *T* is the closest to the center.

And we know the minor radius is **b**, and is parallel to the *y*-axis, that is, perpendicular to the *x*-axis, because the ellipse *E* is horizontal.

So **b** is the distance from *T* to the center when *T* is the closest to the center.

And there are two points where *T* gets positioned when *T* is the closest to the center. The two points are antipodal, too. And the two are: **(0, b)** and **(0, -b)**.

So the minor axis is the line segment connecting the two antipodal points where *T* gets positioned when *T* is the closest to the center, and the two are **(0, b)** and **(0, -b)**.

Thus, the two antipodal points **(0, b)** and **(0, -b)** are the endpoints of the minor axis, and the length of it is **2b**.

36

Examples 3 in Standard Forms

Specify the center and the two main axes of each ellipse below, and then, graph it.

0. $\dfrac{(x+\frac{11}{2})^2}{\frac{25}{4}}+\dfrac{(y-\frac{13}{2})^2}{\frac{9}{4}}=1$ 1. $\dfrac{(x+6)^2}{9}+(y-5)^2=1$ 2. $\dfrac{(x+1)^2}{9}+\dfrac{(y-5)^2}{4}=1$

3. $(x-5)^2+4(y-7)^2=4$ 4. $(x-6)^2+4(y-\frac{7}{2})^2=9$ 5. $4(x+5)^2+9y^2=36$

6. $9x^2+16y^2=144$ 7. $4(x-3)^2+9(y+4)^2=36$ 8. $9(x-7)^2+16(y+\frac{5}{2})^2=36$

9. $9(x-3)^2+25(y+5)^2=225$ A. $(x+4)^2+4(y+\frac{11}{2})^2=9$

Fig. 0

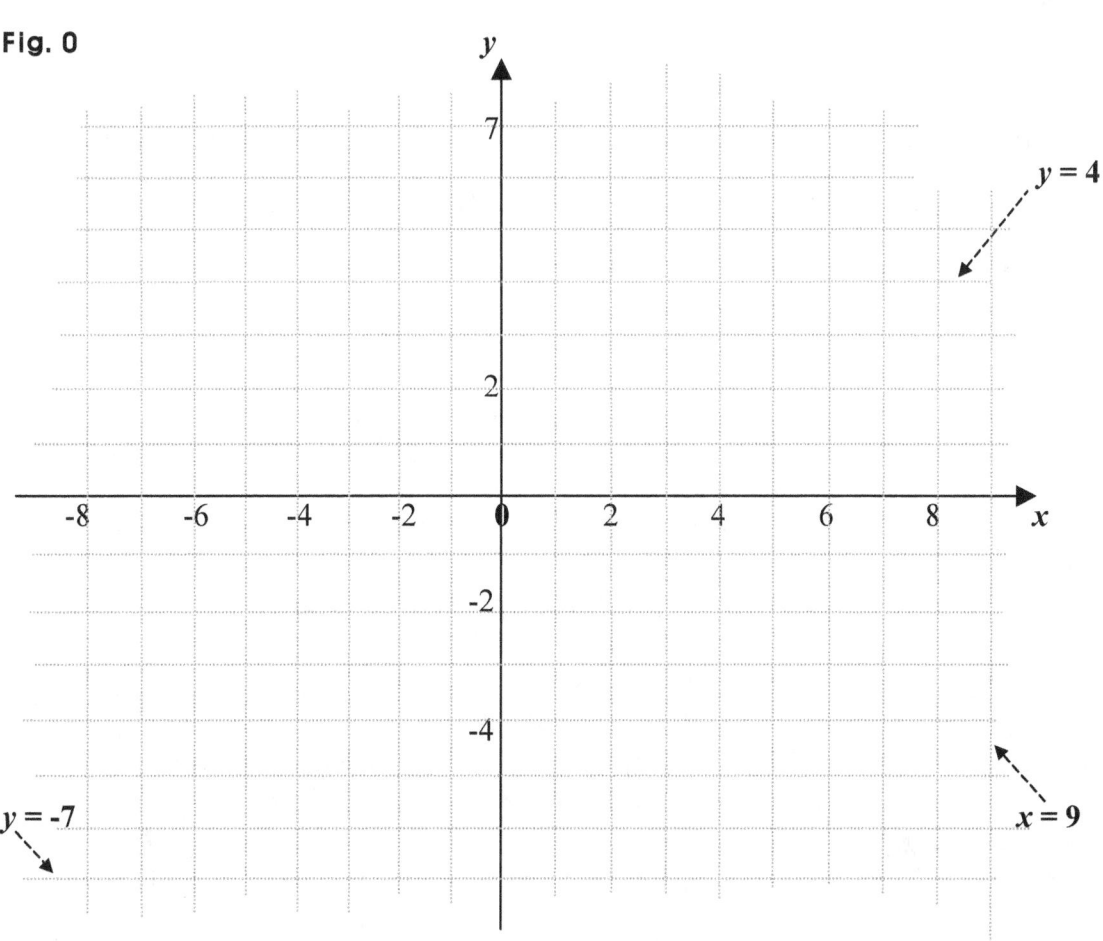

Suggestions or Solutions
To the Problems in the Examples

Specify the center and the major and minor radii of each ellipse below, and then, graph it.

0. $\dfrac{(x+\frac{11}{2})^2}{\frac{25}{4}}+\dfrac{(y-\frac{13}{2})^2}{\frac{9}{4}}=1$ 1. $\dfrac{(x+6)^2}{9}+(y-5)^2=1$ 2. $\dfrac{(x+1)^2}{9}+\dfrac{(y-5)^2}{4}=1$

3. $(x-5)^2+4(y-7)^2=4$ 4. $(x-6)^2+4(y-\frac{7}{2})^2=9$ 5. $4(x+5)^2+9y^2=36$

6. $9x^2+16y^2=144$ 7. $4(x-3)^2+9(y+4)^2=36$ 8. $9(x-7)^2+16(y+\frac{5}{2})^2=36$

9. $9(x-3)^2+25(y+5)^2=225$ A. $(x+4)^2+4(y+\frac{11}{2})^2=9$

First, if an ellipse is **horizontal**, knowing the center and the major and minor radii, we can get the equation of the ellipse using the standard form: $\dfrac{(x-u)^2}{a^2}+\dfrac{(y-v)^2}{b^2}=1,$ where (u, v) is the center, and $\underline{a>b}$, so a is the major radius, and b is the minor radius.

So beginning with 0, we can get: $\dfrac{(x+\frac{11}{2})^2}{\frac{25}{4}}+\dfrac{(y-\frac{13}{2})^2}{\frac{9}{4}}=1 \Rightarrow \dfrac{\{x-(-\frac{11}{2})\}^2}{(\frac{5}{2})^2}+\dfrac{(y-\frac{13}{2})^2}{(\frac{3}{2})^2}=1.$ So we can see that the ellipse is horizontal, the center is at $(-\frac{11}{2},\frac{13}{2})$, the major radius is $\frac{5}{2}$, and the minor radius is $\frac{3}{2}$.

1. $\dfrac{(x+6)^2}{9}+(y-5)^2=1 \Rightarrow \dfrac{\{x-(-6)\}^2}{3^2}+\dfrac{(y-5)^2}{1^2}=1.$ So we can see that the ellipse is horizontal, the center is (-6, 5), the major radius is 3, and the minor radius is 1.

2. $\dfrac{(x+1)^2}{9}+\dfrac{(y-5)^2}{4}=1 \Rightarrow \dfrac{\{x-(-1)\}^2}{3^2}+\dfrac{(y-5)^2}{2^2}=1.$ So we can see that the ellipse is horizontal, the center is (-1, 5), the major radius is 3, and the minor radius is 2.

3. $(x-5)^2 + 4(y-7)^2 = 4 \Rightarrow \dfrac{(x-5)^2}{4} + (y-7)^2 = 1 \Rightarrow \dfrac{(x-5)^2}{2^2} + (y-7)^2 = 1.$

So the ellipse is horizontal, the center is (-1, 5), the major radius is 3, and the minor radius is 2.

4. $(x-6)^2 + 4(y-\frac{7}{2})^2 = 9 \Rightarrow \dfrac{(x-6)^2}{9} + \dfrac{4(y-\frac{7}{2})^2}{9} = 1 \Rightarrow \dfrac{(x-6)^2}{3^2} + \dfrac{(y-\frac{7}{2})^2}{(\frac{3}{2})^2} = 1.$

So the ellipse is horizontal, the center is $(6, \frac{7}{2})$, the major radius is 3, and the minor radius is $\frac{3}{2}$.

5. $4(x+5)^2 + 9y^2 = 36 \Rightarrow \dfrac{4(x+5)^2}{36} + \dfrac{9y^2}{36} = 1 \Rightarrow \dfrac{(x+5)^2}{9} + \dfrac{y^2}{4} = 1 \Rightarrow \dfrac{(x+5)^2}{3^2} + \dfrac{y^2}{2^2} = 1.$

So the ellipse is horizontal, the center is (-5, 0), the major radius is 3, and the minor radius is 2.

6. $9x^2 + 16y^2 = 144 \Rightarrow \dfrac{9x^2}{144} + \dfrac{16y^2}{144} = \dfrac{x^2}{16} + \dfrac{y^2}{9} = \dfrac{x^2}{4^2} + \dfrac{y^2}{3^2} = 1.$

So the ellipse is horizontal, the center is (0, 0), the major radius is 4, and the minor radius is 3.

7. $4(x-3)^2 + 9(y+4)^2 = 36 \Rightarrow \dfrac{(x-3)^2}{9} + \dfrac{(y+4)^2}{4} = \dfrac{(x-3)^2}{3^2} + \dfrac{(y+4)^2}{2^2} = 1.$

So the ellipse is horizontal, the center is (3, -4), the major radius is 3, and the minor radius is 2.

8. $9(x-7)^2 + 16(y+\frac{5}{2})^2 = 36 \Rightarrow \dfrac{(x-7)^2}{4} + \dfrac{4(y+\frac{5}{2})^2}{9} = \dfrac{(x-7)^2}{2^2} + \dfrac{(y+\frac{5}{2})^2}{(\frac{3}{2})^2} = 1.$

So the ellipse is horizontal, the center is $(7, -\frac{5}{2})$, the major radius is 2, and the minor radius is $\frac{3}{2}$.

9. $9(x-3)^2 + 25(y+5)^2 = 225 \Rightarrow \dfrac{(x-3)^2}{25} + \dfrac{(y+5)^2}{9} = \dfrac{(x-3)^2}{5^2} + \dfrac{(y+5)^2}{3^2} = 1.$

So the ellipse is horizontal, the center is (3, -5), the major radius is 5, and the minor radius is 3.

A. $(x+4)^2 + 4(y+\frac{11}{2})^2 = 9 \Rightarrow \dfrac{(x+4)^2}{9} + \dfrac{4(y+\frac{11}{2})^2}{9} = \dfrac{(x+4)^2}{3^2} + \dfrac{(y+\frac{11}{2})^2}{(\frac{3}{2})^2} = 1.$

So the ellipse is horizontal, the center is $(-4, -\frac{11}{2})$, the major radius is 3, and the minor radius is $\frac{3}{2}$.

And graphing the equations, we get:

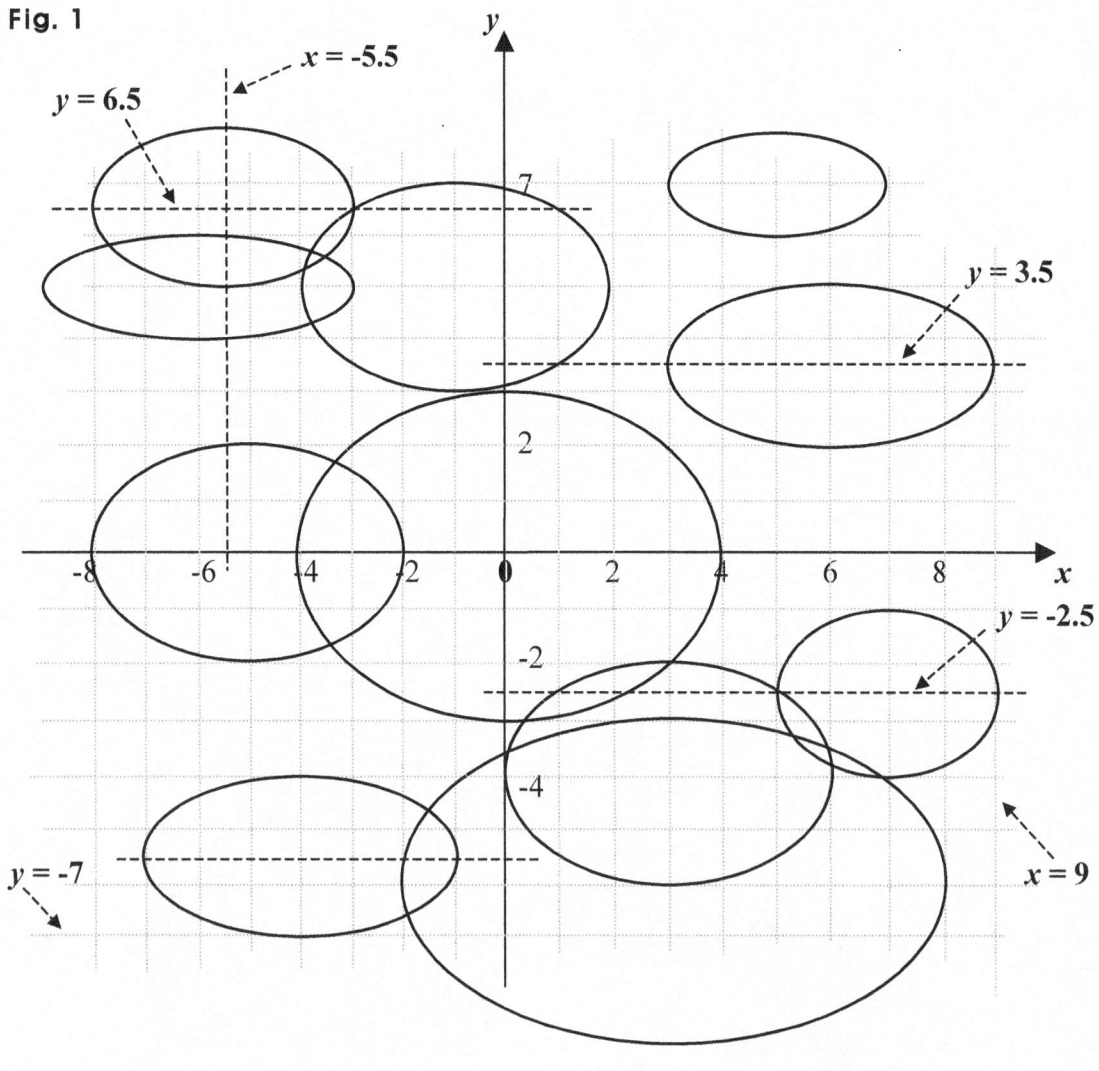

Fig. 1

2. Equations for Ellipses 2

To begin with, putting an ellipse in an equation, we can get:

$$\frac{x^2}{a^2} + \frac{y^2}{b^2} = 1, \text{ where } a > b > 0.$$ What ellipse then, is it?

It is an ellipse <u>horizontal</u>, where the <u>center is at (0, 0)</u>, and its <u>main axes are **2a** and **2b**</u>. So <u>the major axis is parallel to the **x**-axis</u>, and is **2a**, and the ellipse is as below:

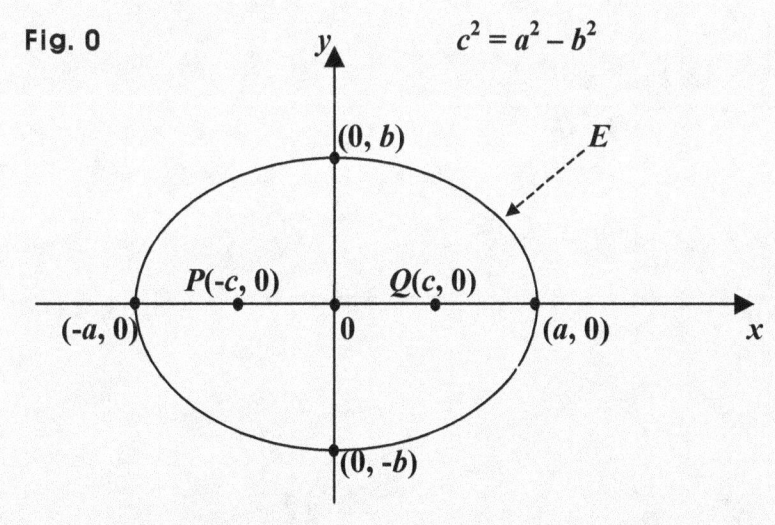

Fig. 0

$c^2 = a^2 - b^2$

What if the ellipse **E** above is centered at a point other than the origin?

For instance, moving the center to (2, 1), we get a new ellipse, so we get a new equation. What then, will the new equation be?

Assuming the new ellipse is **G**, we can get **G** translating the ellipse **E** in the amount of 2 in the direction of the **x**-axis, and in the amount of 1 in the direction of the **y**-axis.

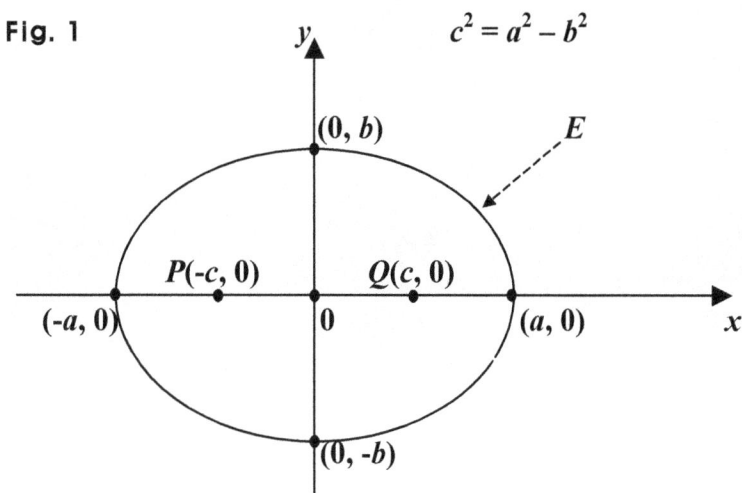

Fig. 1

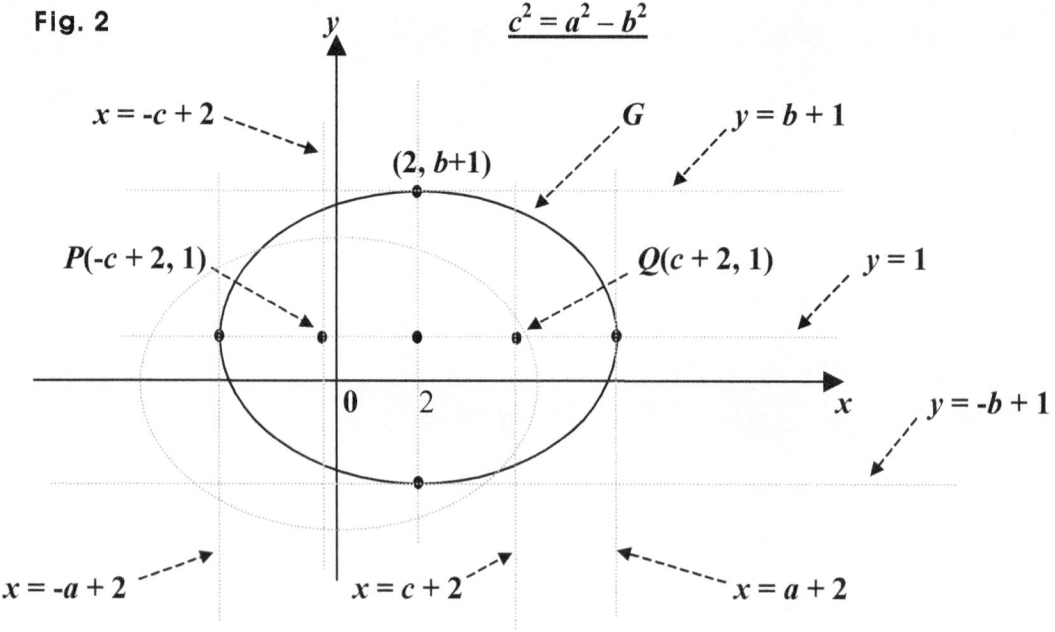

Fig. 2

Then, the equation of the ellipse **G** is: $\dfrac{(x-2)^2}{a^2}+\dfrac{(y-1)^2}{b^2}=1$, where $a > b > 0$.

(Note that we still get: $c^2 = a^2 - b^2$, because **a** and **b** are constant, and so is **c**. And translations do not change shapes, so the focal distance **c** does not change either.)

And in general, we can put a perpendicular ellipse in an equation the way as follows:

$$\frac{(x-u)^2}{a^2} + \frac{(y-v)^2}{b^2} = 1.$$

Then, the ellipse above is centered at a point (u, v), and the main axes are $2a$ and $2b$.

If $a > b > 0$, the ellipse is horizontal, and the major axis is a, and parallel to the x-axis.

If $b > a > 0$, the ellipse is vertical, and the major axis is b, and parallel to the y-axis.

And the equation above is called the standard equation of an ellipse.

(Note however, the standard equation can indicate a perpendicular ellipse only. If an ellipse is perpendicular, one of the main axes is perpendicular to the x-axis.)

What if an ellipse is not perpendicular, thus, slanted as the ellipse below?

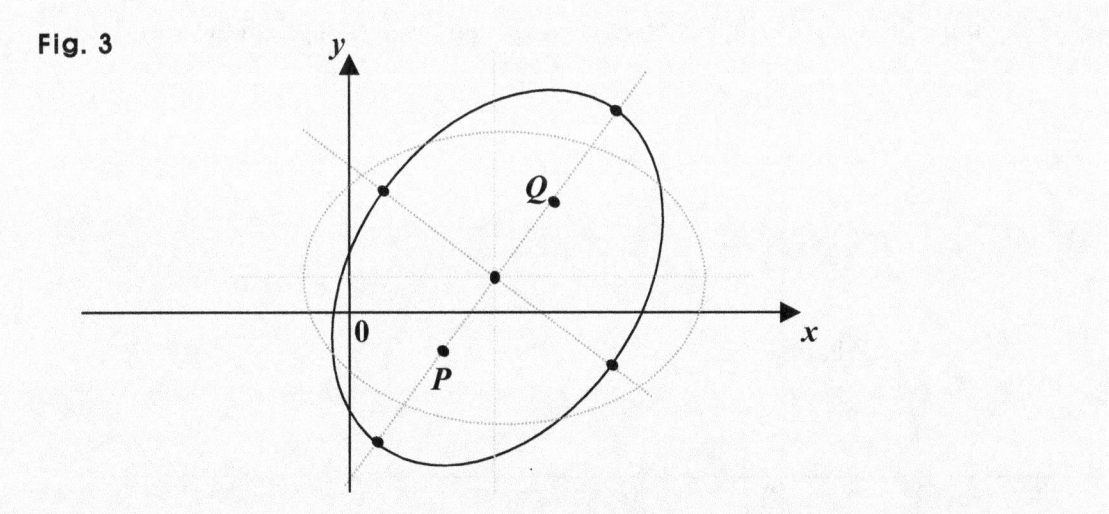

Fig. 3

Given the foci, you can derive the equation using the definition for ellipses, or can get the equation applying a transformation to a perpendicular ellipse.
The transformation is covered in the book, **GRAPH OPERATIONS**, and is covered in the section, Transformation by Turning.

What then, about the ellipse as below?

The center is at (0, 0), its main axes are **2a** and **2b**, and the major axis is this time, parallel to the *y*-axis, and is **2b**.

Then, the ellipse is vertical, and the equation is: $\dfrac{x^2}{a^2} + \dfrac{y^2}{b^2} = 1$, where **b > a > 0**.

So the equation itself looks quite the same as the equation of the horizontal ellipse **E**. We know however, the major > the minor. So the major radius is **b**, because: **b > a > 0**. What then, are the domain and the range?

The domain is still: $-a \le x \le a$, that is, $|x| \le a$, and also, the range is: $-b \le y \le b$ or $|y| \le b$.

And we know the foci are in the major axis, which is this time, parallel to the *y*-axis, and the center is the origin, so if **c** is the focal distance, the foci are **(0, c)** and **(0, -c)**.

Assuming thus, the vertical ellipse above is **V**, we can put **V** in a graph the way below:

Fig. 4

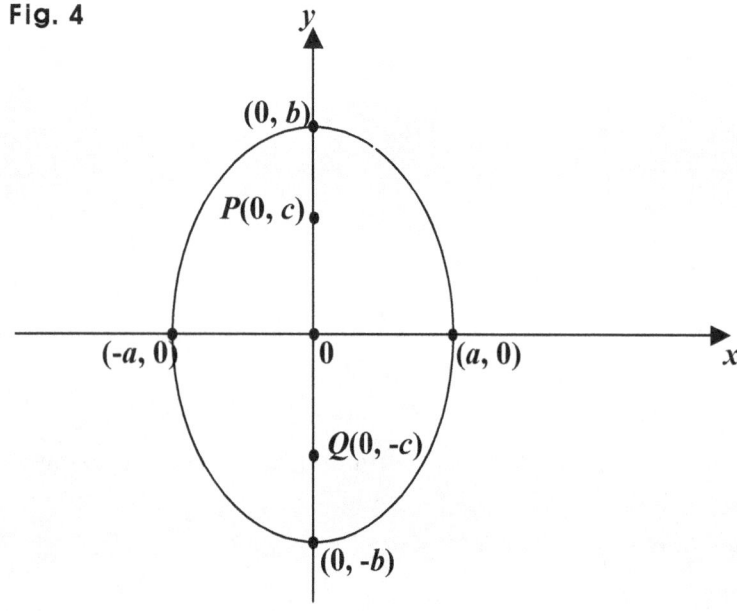

How then, can we get the equation?

We know the definition for ellipses, which says the fact that an ellipse is a set of points, from each of which, the sum of the two distances to two points called the foci is constant.

Using thus, the fact again, we can get the equation of an ellipse vertical, too.

So to begin with, suppose $c > 0$, two points $P(0, c)$ and $Q(0, -c)$ are the two foci of an ellipse called V, and a point $T(x, y)$ is an arbitrary point in the ellipse V, i.e., a random point representing all the points in V.

Then, the center is the origin, that is, $(0, 0)$, and we can say that c is the focal distance, because the center is the midpoint between the two foci. So we can put the three points P, Q, and T in the x-y plane the way as follows:

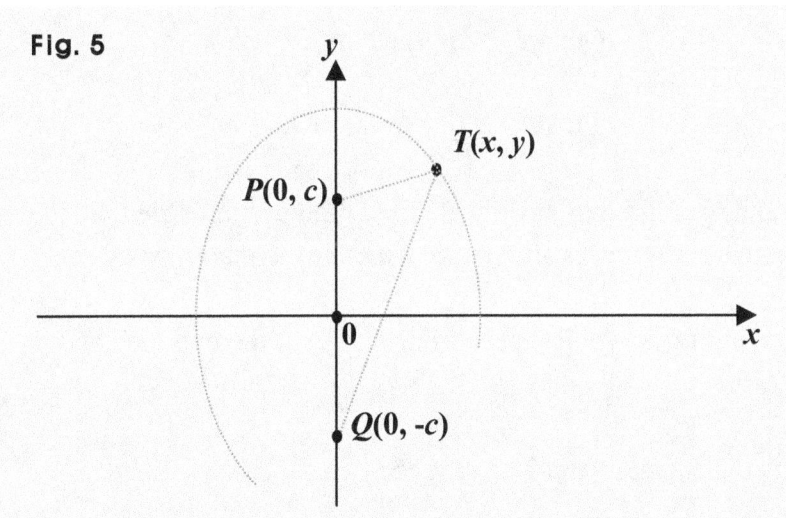

Fig. 5

Suppose next, the point T is moving now, along the ellipse V.

And we know the fact that the sum of the two distances from each point in V to the two foci is constant, that is, the same.   So this time, too, we are going to use the fact again.

So suppose this time, that the point T is now at the point $(0, b)$.

Then, the sum of the two distances is $2b$.   How come?

Fig. 6

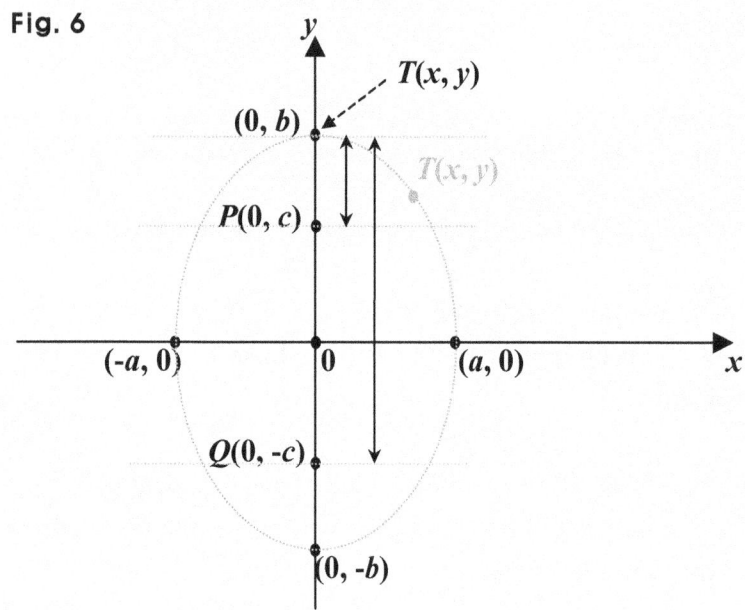

We can see that the length of TQ is: $b - (-c) = b + c$.

And the length of TP is: $b - c$.   So we get: $TP + TQ = (b - c) + (b + c) = 2b$.

Therefore, no matter where the point T may be in the ellipse V, the sum of the two distances from the point T to the two foci is always $2b$, which is the major axis.

So we are going to get the sum of the two distances from $T(x, y)$ to the two foci, and then, set the sum equal to $2b$.

How then, can we get the two distances from $T(x, y)$ to the two foci?

We can use the distance formula, often called Pythagorean Theorem.
And using it, we can get an equation expressed in terms of x and y, that is, we get the connective equation between x and y, which are the coordinates of the arbitrary point $T(x, y)$ in the curve called the ellipse V.   How then, do we call the equation?

The equation explains every point in the ellipse V, that is, it indicates the ellipse V.
So the equation is called the equation of the ellipse V.

So let's now get the equation.   How then, can we apply the distance formula?

Assuming d is the distance between two points, Δx is the difference in x-coordinates, and Δy is the difference in y-coordinates, we can put the formula the way below:

$$d^2 = (\Delta x)^2 + (\Delta y)^2.$$ And we have: $P(0, c)$, $Q(0, -c)$, and $T(x, y)$.

So using the distance formula, and assuming $TP = p$, and $TQ = q$, we can get:

$$p^2 = (x - 0)^2 + (y - c)^2 = x^2 + (y - c)^2, \text{ and } q^2 = (x - 0)^2 + \{y - (-c)\}^2 = x^2 + (y + c)^2.$$

So we get: $TP = \sqrt{x^2 + (y - c)^2}$, and $TQ = \sqrt{x^2 + (y + c)^2}$.

Thus, we get: $TP + TQ = \sqrt{x^2 + (y - c)^2} + \sqrt{x^2 + (y + c)^2} = 2b$. Then, we get:

$$\sqrt{x^2 + (y - c)^2} = 2b - \sqrt{x^2 + (y + c)^2} \Rightarrow x^2 + (y - c)^2 = (2b - \sqrt{x^2 + (y + c)^2})^2$$

$$\Rightarrow x^2 + (y - c)^2 = 4b^2 - 4b\sqrt{x^2 + (y + c)^2} + x^2 + (y + c)^2$$

$$\Rightarrow 4b\sqrt{x^2 + (y + c)^2} = (y + c)^2 - (y - c)^2 + 4b^2 = y^2 + 2cy + c^2 - y^2 + 2cy - c^2 + 4b^2$$

$$\Rightarrow 4b\sqrt{x^2 + (y + c)^2} = 4b^2 + 4cy \Rightarrow b\sqrt{x^2 + (y + c)^2} = b^2 + cy$$

$$\Rightarrow b^2\{x^2 + (y + c)^2\} = (b^2 + cy)^2 = b^4 + 2b^2cy + c^2y^2$$

$$\Rightarrow b^2(x^2 + y^2 + 2cy + c^2) = b^2x^2 + b^2y^2 + 2b^2cy + b^2c^2 = a^4 + 2b^2cy + c^2y^2$$

$$\Rightarrow b^2x^2 + b^2y^2 + b^2c^2 - c^2y^2 = b^4 \Rightarrow b^2x^2 + (b^2 - c^2)y^2 = b^4 - b^2c^2 = b^2(b^2 - c^2)$$

$$\Rightarrow \frac{b^2x^2}{b^2 - c^2} + y^2 = b^2 \Rightarrow \frac{x^2}{b^2 - c^2} + \frac{y^2}{b^2} = 1.$$ What then, about $b^2 - c^2$?

It is constant, so assuming a is constant, we can set: $a^2 = b^2 - c^2$. And assuming $a > 0$, we can put two points $(a, 0)$ and $(-a, 0)$ the way below:

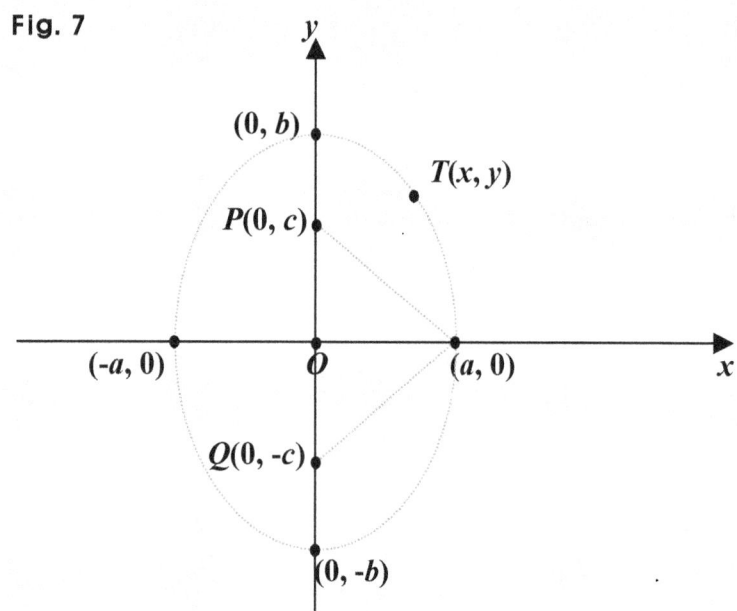

Fig. 7

Assuming now, that the point T is at $(a, 0)$, we can see a triangle isosceles, and the triangle is $\triangle TPQ$.

Then, we can see an isosceles triangle, and the triangle is $\triangle TPQ$.

And we know that the sum of the two distances from T to the two foci is $2b$.

That is, we have: $TP + TQ = 2b$.   So we get: $TP = TQ = b$.

Assuming thus, O is the origin, we can see the triangle TOP is a right triangle, and its hypotenuse TP is b, and the side PO is c.

So applying the distance formula, we can get: $b^2 = a^2 + c^2 \Rightarrow a^2 = b^2 - c^2$.

And we now have: $\dfrac{x^2}{b^2 - c^2} + \dfrac{y^2}{b^2} = 1$.   So we get: $\dfrac{x^2}{a^2} + \dfrac{y^2}{b^2} = 1$,

And we know: $b > a > 0$, since: $a^2 = b^2 - c^2 \Rightarrow b^2 = a^2 + c^2 \Rightarrow b > a > 0$.

And the equation above is expressed in terms of the coordinates of the point $T(x, y)$, so it is an equation in terms of x and y. And we call such an equation a connective equation.

So in this case, the connective equation is the equation that connects the two variables used as the coordinates of the arbitrary point $T(x, y)$ in the curve called the ellipse V.

Thus, the equation of the ellipse V is: $\dfrac{x^2}{a^2} + \dfrac{y^2}{b^2} = 1$, where $\underline{b > a > 0}$.

And putting the ellipse V in the x-y plane, we get:

Fig. 8

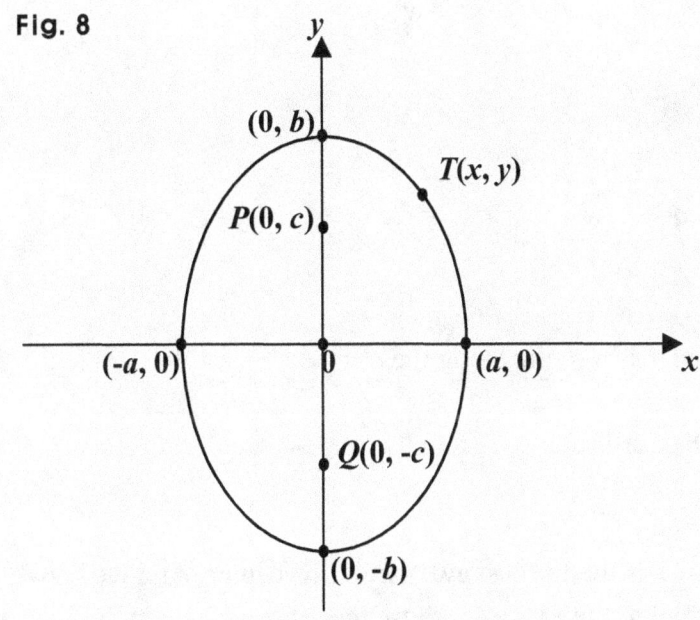

So we can now say that the ellipse V is <u>vertical</u> and centered at the origin, and that the <u>major axis is **2b**</u>, and the minor axis is **2a**.

That is, <u>the semi major axis is **b**</u>, and the semi minor is **a**. In other words, the major radius is **b**, and is <u>parallel to the y-axis</u>, and the minor radius is **a**.

And putting the ellipse V in its equation, we get: $\dfrac{x^2}{a^2} + \dfrac{y^2}{b^2} = 1$, where $\underline{b > a > 0}$.

What then, are the domain and the range?

The domain is $-a \leq x \leq a$, that is, $|x| \leq a$, and the range is: $-b \leq y \leq b$ or $|y| \leq b$.

Fig. 9

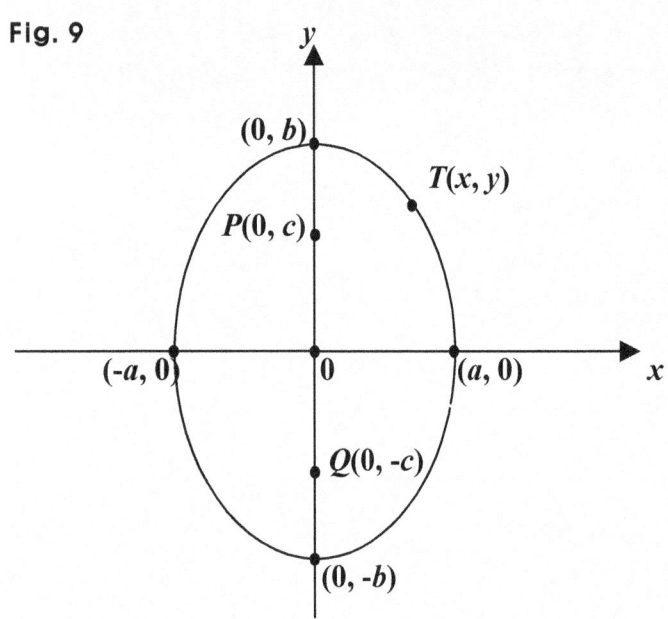

Suppose now again, that the point T is moving along the ellipse V.

Then, when will the point T be the farthest away from the center?

When the point T is in the y-axis, T is the farthest away from the center. And we know the major radius is b, and is parallel to the y-axis, because the ellipse V is vertical.

So b is the distance from T to the center when T is the farthest away from the center.

And the ellipse has two points where T gets located when T is the farthest away from the center. And the two are: $(0, b)$ and $(0, -b)$, which are called antipodal points.

So the major axis is the line segment connecting the two antipodal points where T gets positioned when T is the farthest away from the center, and the two are: $(0, b)$ and $(0, -b)$.

Thus, the two antipodal points $(0, b)$ and $(0, -b)$ are the endpoints of the major axis, and the length of it is **2b**.

What then, about the case where the point T is the closest to the center?

When the point T is in the x-axis, T is the closest to the center.

And we know this time the minor radius is a, and is parallel to the x-axis, that is, perpendicular to the y-axis, because the ellipse V is vertical.

So a is the distance from T to the center when T is the closest to the center.

And there are two points where T gets positioned when T is the closest to the center. The two points are antipodal, too. And the two are: $(a, 0)$ and $(-a, 0)$.

So the minor axis is the line segment connecting the two antipodal points where T gets positioned when T is the closest to the center, and the two are $(a, 0)$ and $(-a, 0)$.

Thus, the two antipodal points $(a, 0)$ and $(-a, 0)$ are the endpoints of the minor axis, and the length of it is $2a$.

So in sum, we have just covered two equations indicating ellipses in two kinds.

One is: $\dfrac{x^2}{a^2} + \dfrac{y^2}{b^2} = 1$, where $\underline{a > b > 0}$, and $\underline{c^2 = a^2 - b^2}$, where c is the focal distance.

The ellipse above is <u>horizontal</u>, and is <u>centered at $(0, 0)$</u>, the <u>major axis is $2a$</u> and is a part of (thus, parallel to) the x-axis, and the minor axis is $2b$.   So the ellipse is as below:

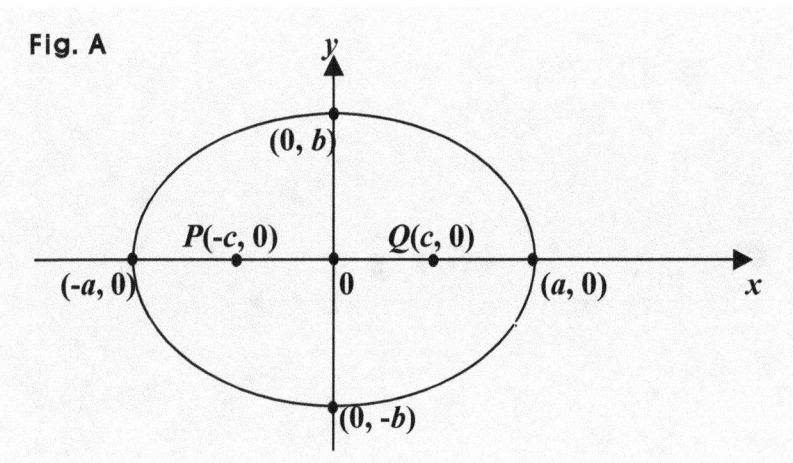

Fig. A

The other is: $\dfrac{x^2}{a^2} + \dfrac{y^2}{b^2} = 1$, where **_b > a > 0_**, **_c² = b² − a²_**, and **_c_** is the focal distance.

The ellipse above is <u>vertical</u>, and is <u>centered at (0, 0)</u>, the <u>major axis is **2b**</u> and is a part of (thus, parallel to) the **_y_**-axis, and the minor axis is **2a**.   So the ellipse is as below:

Fig. B

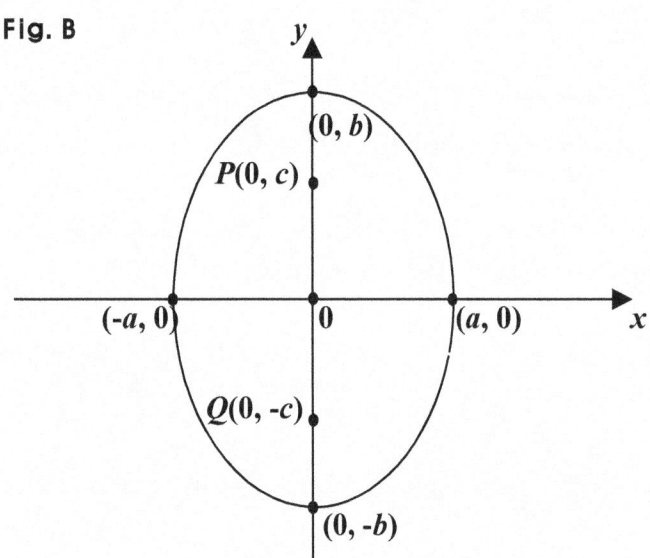

What if the center is not at the origin as in the case below?

Fig. C

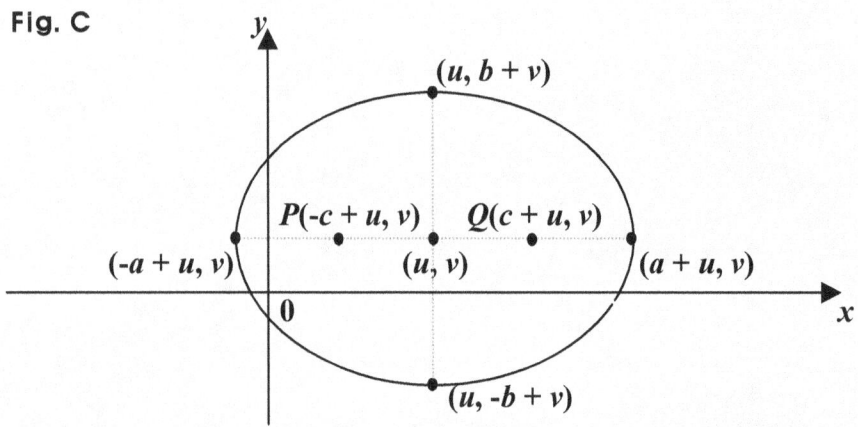

Then, the equation is: $\dfrac{(x-u)^2}{a^2}+\dfrac{(y-v)^2}{b^2}=1$, where $\underline{a>b>0}$, and $\underline{c^2=a^2-b^2}$.

Note that we can get the ellipse above translating an ellipse in amount of u along the x-axis, and in the amount of v along the y-axis. And the ellipse to be translated is as below:

$\dfrac{x^2}{a^2}+\dfrac{y^2}{b^2}=1$, where $\underline{a>b>0}$, and $\underline{c^2=a^2-b^2}$, where c is the focal distance.

And in the case below:

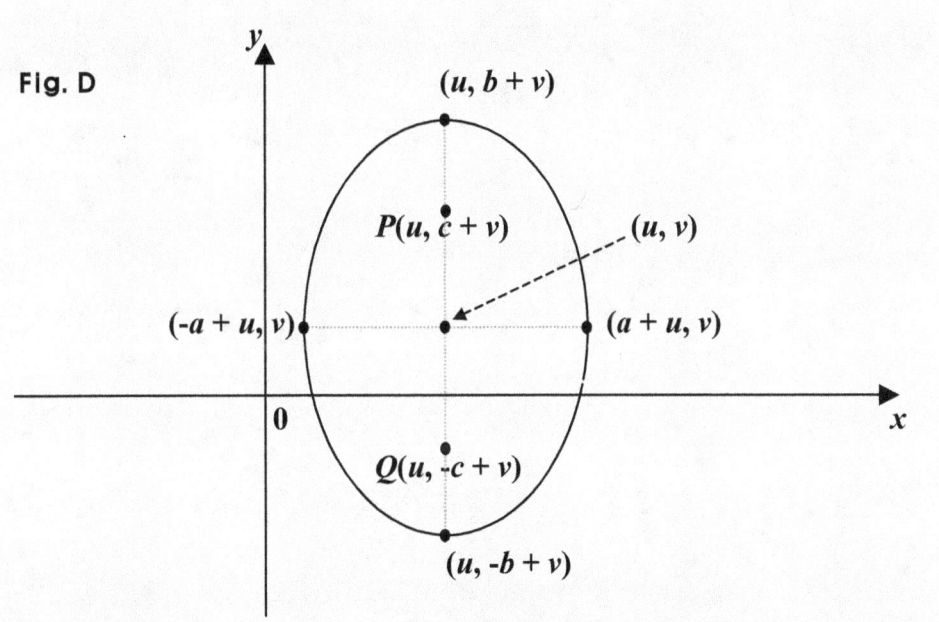

Fig. D

The equation is: $\dfrac{(x-u)^2}{a^2}+\dfrac{(y-v)^2}{b^2}=1$, where $\underline{b>a>0}$, and $\underline{c^2=b^2-a^2}$.

And note that in this case, we have this: $\underline{c^2=b^2-a^2}$, and not this: $c^2=a^2-b^2$.

And also, assuming K is the ellipse above, we can get K translating an ellipse in amount of u along the x-axis, and in the amount of v along the y-axis. And the ellipse to be translated is as below:

$\dfrac{x^2}{a^2} + \dfrac{y^2}{b^2} = 1$, where $\underline{b > a > 0}$, and $\underline{c^2 = b^2 - a^2}$, where c is the focal distance.

And also, assuming L is the ellipse above, we can get L translating the ellipse K in amount of $-u$ along the x-axis, and in the amount of $-v$ along the y-axis.

Examples 4 in Standard Forms

Specify the center and the two main axes of each ellipse below, and then, graph it.

0. $\dfrac{(x+\frac{13}{2})^2}{\frac{9}{4}}+\dfrac{(y-\frac{13}{2})^2}{\frac{25}{4}}=1$ 1. $\dfrac{(x+5)^2}{4}+\dfrac{(y-\frac{11}{2})^2}{\frac{25}{4}}=1$ 2. $\dfrac{x^2}{4}+\dfrac{(y-5)^2}{9}=1$

3. $4(x-\frac{9}{2})^2+(y-7)^2=9$ 4. $196(x-7)^2+100(y-\frac{7}{2})^2=1225$

5. $\dfrac{(x+\frac{9}{2})^2}{25}+\dfrac{(y-\frac{1}{2})^2}{49}=\dfrac{1}{4}$ 6. $25x^2+9y^2=225$ 7. $49(x-3)^2+36(y-\frac{5}{2})^2=441$

8. $\dfrac{(x-5)^2}{9}+\dfrac{4(y+\frac{9}{2})^2}{81}=1.$ 9. $\dfrac{(x-7)^2}{16}+\dfrac{(y+\frac{5}{2})^2}{25}=\dfrac{1}{4}$ A. $\dfrac{(x+4)^2}{36}+\dfrac{(y+\frac{11}{2})^2}{49}=\dfrac{1}{4}$

Fig. 0

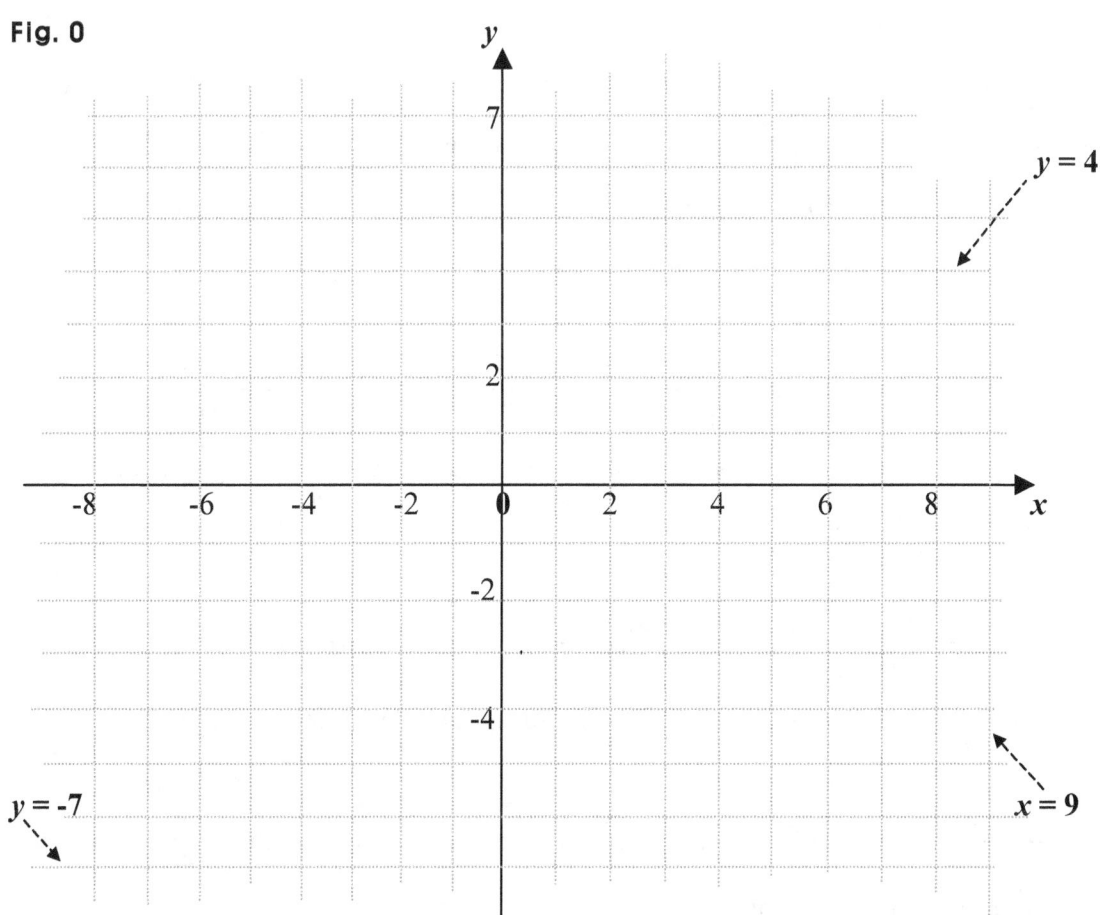

56

Suggestions or Solutions
To the Problems in the Examples

Specify the center and the major and minor radii of each ellipse below, and then, graph it.

0. $\dfrac{(x+\frac{13}{2})^2}{\frac{9}{4}}+\dfrac{(y-\frac{13}{2})^2}{\frac{25}{4}}=1$ 1. $\dfrac{(x+5)^2}{4}+\dfrac{(y-\frac{11}{2})^2}{\frac{25}{4}}=1$ 2. $\dfrac{x^2}{4}+\dfrac{(y-5)^2}{9}=1$

3. $4(x-\frac{9}{2})^2+(y-7)^2=9$ 4. $196(x-7)^2+100(y-\frac{7}{2})^2=1225$

5. $\dfrac{(x+\frac{9}{2})^2}{25}+\dfrac{(y-\frac{1}{2})^2}{49}=\dfrac{1}{4}$ 6. $25x^2+9y^2=225$ 7. $49(x-3)^2+36(y-\frac{5}{2})^2=441$

8. $\dfrac{(x-5)^2}{9}+\dfrac{4(y+\frac{9}{2})^2}{81}=1.$ 9. $\dfrac{(x-7)^2}{16}+\dfrac{(y+\frac{5}{2})^2}{25}=\dfrac{1}{4}$ A. $\dfrac{(x+4)^2}{36}+\dfrac{(y+\frac{11}{2})^2}{49}=\dfrac{1}{4}$

First, if an ellipse is **vertical**, knowing the center and the major and minor radii, we can get the equation of the ellipse using the standard form: $\dfrac{(x-u)^2}{a^2}+\dfrac{(y-v)^2}{b^2}=1,$ where (u, v) is the center, and $\underline{b > a}$, so b is the major radius, and a is the minor radius.

So beginning with 0, we can get: $\dfrac{(x+\frac{13}{2})^2}{\frac{9}{4}}+\dfrac{(y-\frac{13}{2})^2}{\frac{25}{4}}=1\Rightarrow\dfrac{(x+\frac{13}{2})^2}{(\frac{3}{2})^2}+\dfrac{(y-\frac{13}{2})^2}{(\frac{5}{2})^2}=1.$
So we can see that the ellipse is vertical, the center is at $(-\frac{13}{2},\frac{13}{2})$, the major radius is $\frac{5}{2}$, and the minor radius is $\frac{3}{2}$.

1. $\dfrac{(x+5)^2}{4}+\dfrac{(y-\frac{11}{2})^2}{\frac{25}{4}}=1\Rightarrow\dfrac{\{x-(-5)\}^2}{2^2}+\dfrac{(y-\frac{11}{2})^2}{(\frac{5}{2})^2}=1.$ So we can see that the ellipse is vertical, the center is $(-5,\frac{11}{2})$, the major radius is $\frac{5}{2}$, and the minor radius is 2.

2. $\dfrac{x^2}{4}+\dfrac{(y-5)^2}{9}=1.\Rightarrow\dfrac{(x-0)^2}{2^2}+\dfrac{(y-5)^2}{3^2}=1.$ So we can see that the ellipse is vertical, the center is (0, 5), the major radius is 3, and the minor radius is 2.

3. $4(x-\frac{9}{2})^2+(y-7)^2=9 \Rightarrow \frac{4(x-\frac{9}{2})^2}{9}+\frac{(y-7)^2}{9}=1 \Rightarrow \frac{(x-\frac{9}{2})^2}{(\frac{3}{2})^2}+\frac{(y-7)^2}{3^2}=1.$ So the

ellipse is vertical, the center is $(\frac{9}{2},7)$, the major radius is 3, and the minor radius is $\frac{3}{2}$.

4. $196(x-7)^2+100(y-\frac{7}{2})^2=1225 \Rightarrow \frac{4(x-7)^2}{25}+\frac{4(y-\frac{7}{2})^2}{49}=1 \Rightarrow \frac{(x-7)^2}{(\frac{5}{2})^2}+\frac{(y-\frac{7}{2})^2}{(\frac{7}{2})^2}=1.$

So the ellipse is vertical, the center is $(7,\frac{7}{2})$, the major radius is $\frac{7}{2}$, and the minor radius

is $\frac{5}{2}$.

5. $\frac{(x+\frac{9}{2})^2}{25}+\frac{(y-\frac{1}{2})^2}{49}=\frac{1}{4} \Rightarrow \frac{4(x+\frac{9}{2})^2}{25}+\frac{4(y-\frac{1}{2})^2}{49}=1 \Rightarrow \frac{(x+\frac{9}{2})^2}{(\frac{5}{2})^2}+\frac{(y-\frac{1}{2})^2}{(\frac{7}{2})^2}=1.$

So the ellipse is vertical, the center is $(-\frac{9}{2},\frac{1}{2})$, the major radius is $\frac{7}{2}$, and the minor radius

is $\frac{5}{2}$.

6. $25x^2+9y^2=225 \Rightarrow \frac{25x^2}{225}+\frac{9y^2}{225}=\frac{x^2}{9}+\frac{y^2}{25}=\frac{x^2}{3^2}+\frac{y^2}{5^2}=1.$

So the ellipse is vertical, the center is (0, 0), the major radius is 5, and the minor radius is
3.

7. $4(x-3)^2+9(y+4)^2=36 \Rightarrow \frac{(x-3)^2}{9}+\frac{(y+4)^2}{4}=\frac{(x-3)^2}{3^2}+\frac{(y+4)^2}{2^2}=1.$ So the

ellipse is vertical, the center is (3, -4), the major radius is 3, and the minor radius is 2.

8. $\frac{(x-5)^2}{9}+\frac{4(y+\frac{9}{2})^2}{81}=1 \Rightarrow \frac{(x-5)^2}{9}+\frac{(y+\frac{9}{2})^2}{\frac{81}{4}}=\frac{(x-5)^2}{3^2}+\frac{(y+\frac{9}{2})^2}{(\frac{9}{2})^2}=1.$ So the ellipse

is vertical, the center is $(5,-\frac{9}{2})$, the major radius is $\frac{9}{2}$, and the minor radius is 3.

9. $\frac{(x-7)^2}{16}+\frac{(y+\frac{5}{2})^2}{25}=\frac{1}{4} \Rightarrow \frac{4(x-7)^2}{16}+\frac{4(y+\frac{5}{2})^2}{25}=1 \Rightarrow \frac{(x-7)^2}{2^2}+\frac{(y+\frac{5}{2})^2}{(\frac{5}{2})^2}=1.$

So the ellipse is vertical, the center is $(7,-\frac{5}{2})$, the major radius is $\frac{5}{2}$, and the minor radius

is 2.

A. $\dfrac{(x+4)^2}{36}+\dfrac{(y+\frac{11}{2})^2}{49}=\dfrac{1}{4} \Rightarrow \dfrac{4(x+4)^2}{36}+\dfrac{4(y+\frac{11}{2})^2}{49}=1 \Rightarrow \dfrac{(x+4)^2}{3^2}+\dfrac{(y+\frac{11}{2})^2}{(\frac{7}{2})^2}=1.$

So the ellipse is vertical, the center is $(-4,-\frac{11}{2})$, the major radius is $\frac{7}{2}$, and the minor radius is 3.

And graphing the equations, we get:

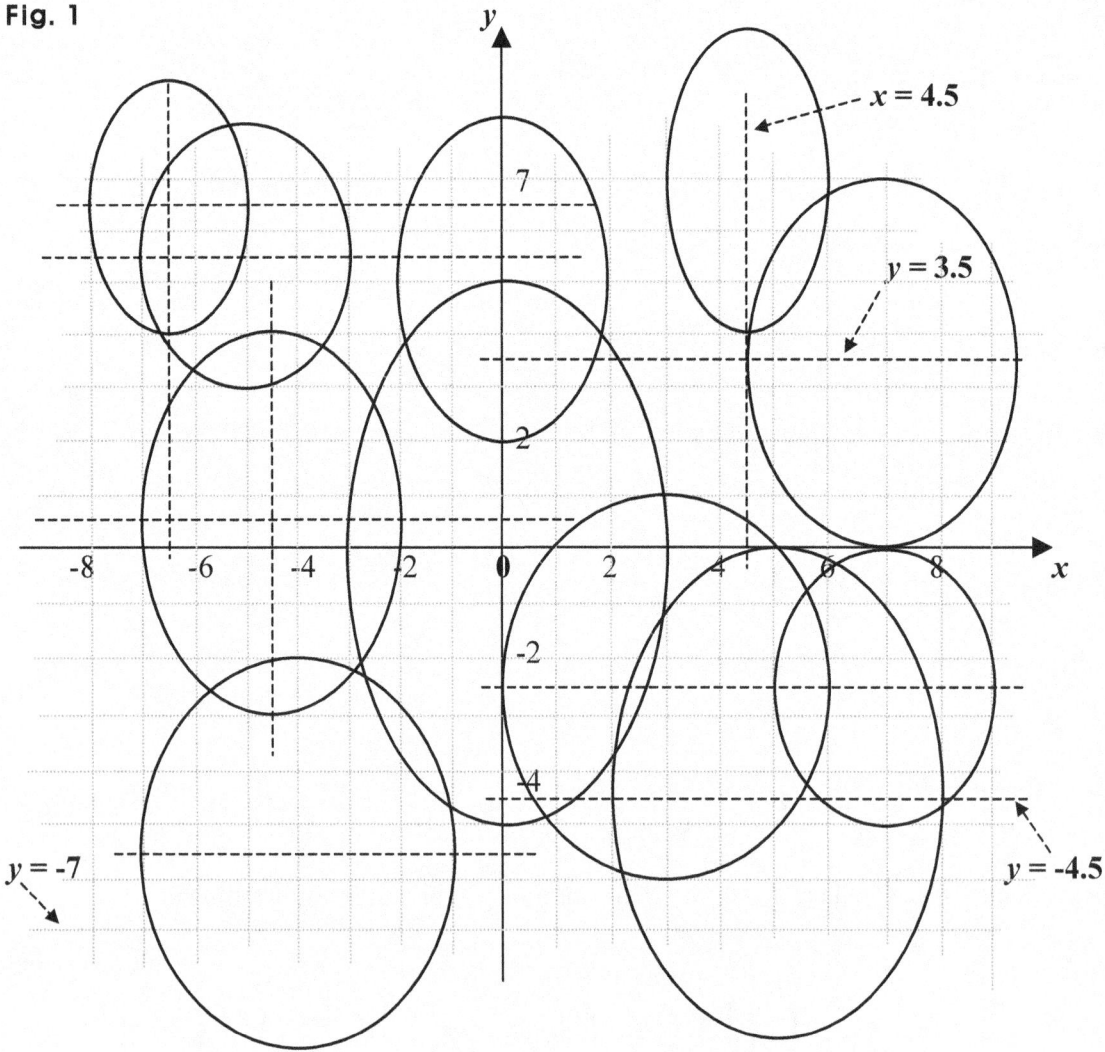

Examples 5 in Standard Forms

Find the center and the main axes of each ellipse below:

0. $x^2 + 2x + 2y^2 + 2y - 1 = 0$

1. $36x^2 + 36x + 100y^2 - 300y + 9 = 0$

2. $x^2 + 12x + 9y^2 - 90y + 252 = 0$

3. $2x - 2x^2 + 3y - y^2 + 1 = 0$

4. $25x^2 + 250x + 16y^2 + 16y + 529 = 0$

5. $9x^2 + 4y^2 - 24y = 0$

Suggestions or Solutions
To the Problems in the Examples

0. $x^2 + 2x + 2y^2 + 2y - 1 = 0 \Rightarrow x^2 + 2x + 1 + 2(y^2 + y) - 1 = 0$

$\Rightarrow (x+1)^2 + 2\{y^2 + y + (\frac{1}{2})^2 - (\frac{1}{2})^2\} - 1 = 0 \Rightarrow (x+1)^2 + 2\{y^2 + y + (\frac{1}{2})^2\} - \frac{1}{2} - 1 = 0$

$\Rightarrow (x+1)^2 + 2(y+\frac{1}{2})^2 - \frac{3}{2} = 0 \Rightarrow (x+1)^2 + 2(y+\frac{1}{2})^2 = \frac{3}{2} \Rightarrow \dfrac{(x+1)^2}{\frac{3}{2}} + \dfrac{2(y+\frac{1}{2})^2}{\frac{3}{2}} = 1$

$\Rightarrow \dfrac{(x+1)^2}{(\sqrt{\frac{3}{2}})^2} + \dfrac{(y+\frac{1}{2})^2}{\frac{3}{4}} = 1 \Rightarrow \dfrac{(x+1)^2}{(\sqrt{\frac{3}{2}})^2} + \dfrac{(y+\frac{1}{2})^2}{(\frac{\sqrt{3}}{2})^2} = 1.$

So the ellipse is horizontal, the center is $(-1, -\frac{1}{2})$, the major radius is $\sqrt{\frac{3}{2}}$, and the minor radius is $\frac{\sqrt{3}}{2}$.

1. $36x^2 + 36x + 100y^2 - 300y + 9 = 0 \Rightarrow 36(x^2 + x) + 100(y^2 - 3y) + 9 = 0$

$\Rightarrow 36(x^2 + x + \frac{1}{4} - \frac{1}{4}) + 100(y^2 - 3y + \frac{9}{4} - \frac{9}{4}) + 9 = 0$

$\Rightarrow 36(x^2 + x + \frac{1}{4}) - 9 + 100(y^2 - 3y + \frac{9}{4}) - 225 + 9 = 0$

$\Rightarrow 36(x + \frac{1}{2})^2 + 100(y - \frac{3}{2})^2 - 225 = 0 \Rightarrow 36(x + \frac{1}{2})^2 + 100(y - \frac{3}{2})^2 = 225$

$\Rightarrow \dfrac{36(x+\frac{1}{2})^2}{225} + \dfrac{100(y-\frac{3}{2})^2}{225} = 1 \Rightarrow \dfrac{4(x+\frac{1}{2})^2}{25} + \dfrac{4(y-\frac{3}{2})^2}{9} = 1 \Rightarrow \dfrac{(x+\frac{1}{2})^2}{\frac{25}{4}} + \dfrac{(y-\frac{3}{2})^2}{\frac{9}{4}} = 1$

$\Rightarrow \dfrac{(x+\frac{1}{2})^2}{(\frac{5}{2})^2} + \dfrac{(y-\frac{3}{2})^2}{(\frac{3}{2})^2} = 1.$

So the ellipse is horizontal, the center is $(-\frac{1}{2}, \frac{3}{2})$, the major radius is $\frac{5}{2}$, and the minor radius is $\frac{3}{2}$.

2. $x^2 + 12x + 9y^2 - 90y + 252 = 0$

$\Rightarrow x^2 + 12x + 9y^2 - 90y + 252 = x^2 + 12x + 9(y^2 - 10y) + 252$

$= x^2 + 12x + 36 - 36 + 9(y^2 - 10y + 25 - 25) + 252$

$= (x + 6)^2 - 36 + 9(y^2 - 10y + 25) - 225 + 252$

$= (x + 6)^2 + 9(y - 5)^2 - 9 = 0$

$\Rightarrow \dfrac{(x+6)^2}{9} + (y-5)^2 = 1 \Rightarrow \dfrac{\{x-(-6)\}^2}{3^2} + \dfrac{(y-5)^2}{1^2} = 1.$

So the ellipse is horizontal, the center is (-6, 5), the major radius is 3, and the minor radius is 1.

3. $2x - 2x^2 + 3y - y^2 + 1 = 0 \Rightarrow 2x^2 - 2x + y^2 - 3y - 1 = 0$

$\Rightarrow 2x^2 - 2x + y^2 - 3y - 1 = 2(x^2 + x + \tfrac{1}{4} - \tfrac{1}{4}) + (y^2 - 3y + \tfrac{9}{4} - \tfrac{9}{4}) - 1$

$= 2(x + \tfrac{1}{2})^2 - \tfrac{1}{2} + (y - \tfrac{3}{2})^2 - \tfrac{9}{4} - 1 = 2(x + \tfrac{1}{2})^2 + (y - \tfrac{3}{2})^2 - \tfrac{15}{4} = 0$

$\Rightarrow 2(x + \tfrac{1}{2})^2 + (y - \tfrac{3}{2})^2 = \tfrac{15}{4} \Rightarrow \dfrac{2(x+\frac{1}{2})^2}{\frac{15}{4}} + \dfrac{(y-\frac{3}{2})^2}{\frac{15}{4}} = 1 \Rightarrow \dfrac{(x+\frac{1}{2})^2}{\frac{15}{8}} + \dfrac{(y-\frac{3}{2})^2}{\frac{15}{4}} = 1$

$\Rightarrow \dfrac{(x+\frac{1}{2})^2}{(\frac{\sqrt{15}}{2\sqrt{2}})^2} + \dfrac{(y-\frac{3}{2})^2}{(\frac{\sqrt{15}}{2})^2} = 1 \Rightarrow \dfrac{(x+\frac{1}{2})^2}{(\frac{\sqrt{30}}{4})^2} + \dfrac{(y-\frac{3}{2})^2}{(\frac{\sqrt{15}}{2})^2} = 1.$

So the ellipse is vertical, the center is $(-\tfrac{1}{2}, \tfrac{3}{2})$, the major radius is $\tfrac{\sqrt{15}}{2}$, and the minor radius is $\tfrac{\sqrt{30}}{4}$.

4. $25x^2 + 250x + 16y^2 + 16y + 529 = 0$

$\Rightarrow 25x^2 + 250x + 16y^2 - 16y + 529 = 25(x^2 + 10x) + 16(y^2 - y) + 529$

$= 25(x^2 + 10x + 25 - 25) + 16(y^2 - y + \frac{1}{4} - \frac{1}{4}) + 529$

$= 25(x^2 + 10x + 25) - 625 + 16(y^2 - y + \frac{1}{4}) - 4 + 529$

$= 25(x + 5)^2 + 16(y - \frac{1}{2})^2 - 100 = 0 \Rightarrow 25(x + 5)^2 + 16(y - \frac{1}{2})^2 = 100$

$\Rightarrow \dfrac{25(x+5)^2}{100} + \dfrac{16(y-\frac{1}{2})^2}{100} = 1 \Rightarrow \dfrac{(x+5)^2}{4} + \dfrac{4(y-\frac{1}{2})^2}{25} = 1 \Rightarrow \dfrac{(x+5)^2}{4} + \dfrac{(y-\frac{1}{2})^2}{\frac{25}{4}} = 1$

$\Rightarrow \dfrac{\{x-(-5)\}^2}{2^2} + \dfrac{(y-\frac{1}{2})^2}{(\frac{5}{2})^2} = 1.$

So the ellipse is vertical, the center is $(-5, \frac{1}{2})$, the major radius is $\frac{5}{2}$, and the minor radius is 2.

5. $9x^2 + 4y^2 - 24y = 0$

$\Rightarrow 9x^2 + 4y^2 - 24y = 9x^2 + 4(y^2 - 6y) = 9x^2 + 4(y^2 - 6y + 9 - 9)$

$= 9x^2 + 4(y^2 - 6y + 9) - 36 = 9x^2 + 4(y - 3)^2 - 36 = 0 \Rightarrow 9x^2 + 4(y - 3)^2 = 36$

$\Rightarrow \dfrac{9x^2}{36} + \dfrac{4(y-3)^2}{36} = 1 \Rightarrow \dfrac{x^2}{4} + \dfrac{(y-3)^2}{9} = 1 \Rightarrow \dfrac{x^2}{2^2} + \dfrac{(y-3)^2}{3^2} = 1.$

So the ellipse is vertical, the center is $(0, 3)$, the major radius is 3, and the minor radius is 2.

₃.**Elements of an Ellipse**

Working with an ellipse, we often need to refer to some basic elements in it. And the elements are as follows:

The center, foci, directrices, vertices, main axes, focal distance, and eccentricity.

So let's now take a close look at each of those elements.

To begin with, an ellipse has <u>two vertices</u>, which are at <u>both ends of the major axis</u>.

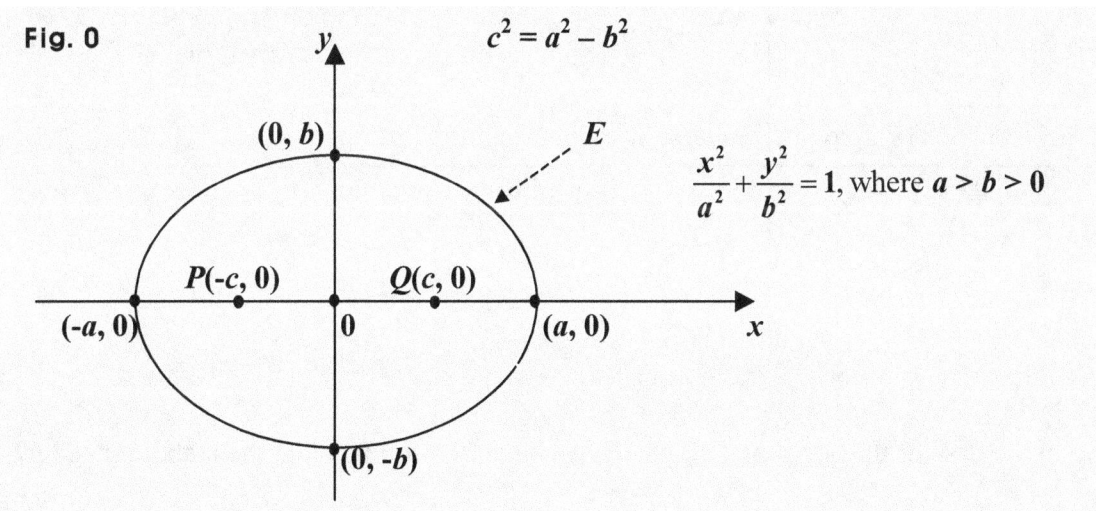

Fig. 0

$$c^2 = a^2 - b^2$$

$$\frac{x^2}{a^2} + \frac{y^2}{b^2} = 1, \text{ where } a > b > 0$$

So in the ellipse E above, the two vertices are **(-a, 0)** and **(a, 0)**.   What about P and Q?

The two points **P(-c, 0)** and **Q(c, 0)** are the foci, and the center is **(0, 0)**.

And we can notice that has the <u>major axis has all the five points</u>: the two vertices, the two foci, and the center.

And also we can notice that if an ellipse is <u>horizontal</u>, <u>the y-coordinates are the same</u> at all the five points, which share thus, the same y-coordinate.

Then, we can notice also that if an ellipse is <u>vertical</u>, the **x-coordinates are the same** at all the five points: the vertices, foci, and center, which share thus, the same **x**-coordinate.

• Next, an ellipse is symmetric, and has two axes of symmetry, often called main axes.

And an ellipse has two main axes, one is called the major axis, longer than the other, called the minor axis. In the ellipse **E**, the line segment connecting the vertices is the major axis, the length of which is **2a**, and the line segment connecting two points **(0, b)** and **(0, b)** is the minor axis, the length of which is **2b**. Usually though, we just call **2a** the major axis, and call **2b** the minor axis.

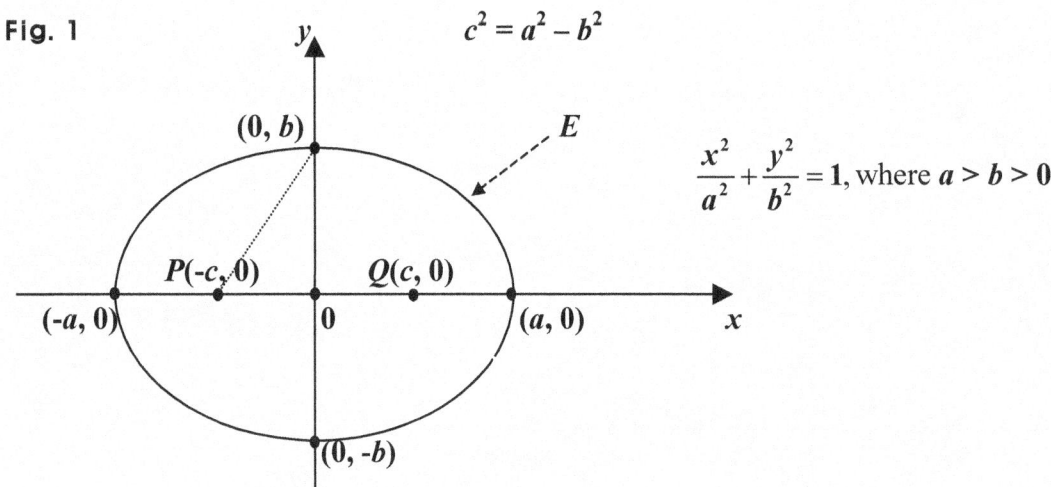

Fig. 1

$c^2 = a^2 - b^2$

$\dfrac{x^2}{a^2} + \dfrac{y^2}{b^2} = 1$, where $a > b > 0$

So in the case of the <u>horizontal</u> ellipse **E**, we have: **a > b > 0**. What then, about **a** and **b**?

We call **a** the semi major axis, called the major radius, too, and **b** is called the semi minor axis, called the minor radius, too.

And as we can see in the equation above, if an ellipse is <u>horizontal</u>, <u>the major radius is with x^2</u>, and is **a**.   What then, about an ellipse vertical?

If the equation above indicates an ellipse <u>vertical</u>, we get: **b > a > 0**.
So <u>the major radius is with y^2</u>, and is **b**.

- Next, the focal distance is the distance from a focus to the center.

The two foci are symmetric about the center.
So the focal distance is half the distance between the foci.
So in the ellipses E above, c is the focal distance.

- Next, in the case of the ellipse E, the two foci are $(-c, 0)$ and $(c, 0)$.

The sum of two distances from every point in an ellipse to its two foci is constant, that is, the same. And in the case of the ellipse E above, the sum is $2a$, the major axis.

What then, about this: $c^2 = a^2 - b^2$?

It can be called the connective equation between the focal distance and the semi axes. And in the case of the ellipse E, we can get the equation from the three facts as follows:

- One is that c is the focal distance, that is, the distance from one focus to the center.

- Another is that the distance from one end of the minor axis to one focus is the major radius, that is, a.

- And the other fact is that a right triangle is made by the three points stated above, one is the one end of the minor axis, another is the one focus, and the other is the center.

And in the case of the ellipse E, the three points can be: $(0, b)$, $(-c, 0)$, and $(0, 0)$.

So the three points above determine a right triangle, where the hypotenuse is a, which is the major radius. Using thus, the distance formula, we can get: $a^2 = b^2 + c^2$.

So we get: $c^2 = a^2 - b^2$.   Therefore, the focal distance is: $c = \sqrt{a^2 - b^2}$.

And in the case of the ellipse E horizontal, the two foci are: $(-c, 0)$ and $(c, 0)$.

So we can put them this way, too: $(-\sqrt{a^2 - b^2},\, 0)$ and $(\sqrt{a^2 - b^2},\, 0)$.

What if the ellipse is vertical?

Then, we have: $b > a > 0$.   So we get: $c^2 = b^2 - a^2$.

Therefore, in the case of an ellipse <u>vertical</u>, the focal distance is: $c = \sqrt{b^2 - a^2}$.

And if an ellipse is <u>vertical and centered at the origin</u>, the two foci are: **(0, *c*)** and **(0, -*c*)**.

So we can put the foci this way, too: **(0, $\sqrt{b^2 - a^2}$)** and **(0, $-\sqrt{b^2 - a^2}$)**.

• Next, an ellipse has a special number called an <u>eccentricity</u>, which is a ratio of a focal distance to a major radius, that is, <u>the focal distance over the major radius</u>.

Fig. 2 $c^2 = a^2 - b^2$, and $e = c/a$

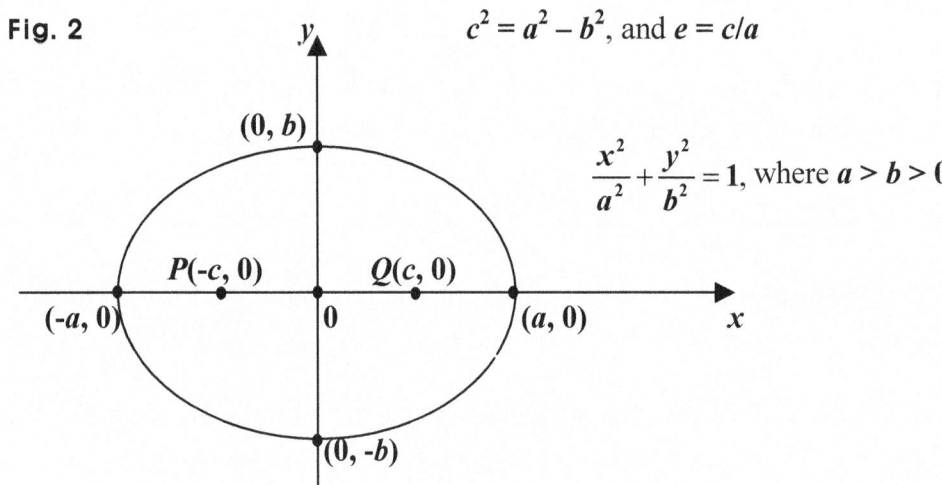

$\dfrac{x^2}{a^2} + \dfrac{y^2}{b^2} = 1$, where $a > b > 0$

So of an ellipse, the eccentricity specifies the degree, to which the ellipse is out of round. Thus, it can tell us how flat or round an ellipse is.

Usually, an eccentricity is denoted by *e*, and we have: **0 < *e* < 1** for an ellipse.

And if the eccentricity *e* were 0, it would be a circle.

If *e* is close to 1, the ellipse is very flat, and looks like a bar.

So the bigger the eccentricity, the flatter the ellipse.   What number though, is *e*?

The eccentricity *e* is a ratio, which is <u>the focal distance over the major radius</u>.

So in the case of an ellipse <u>horizontal</u>, we get: $\underline{e = c/a}$, since $\underline{a > b > 0}$.

And in the case of an ellipse vertical, we get: $e = c/b$, since $b > a > 0$.

So we can put the focal distance c the ways below, too:

If $a > b > 0$, since: $e = c/a$, we get: $c = ae$, and if $b > a > 0$, since: $e = c/b$, we get: $c = be$.

How come though, it would be a circle if it were the case where $e = 0$?

If $a > b > 0$, we have: $e = c/a$, and if $b > a > 0$, we have: $e = c/b$.

So either way, we get: $c = 0$ if $e = 0$.

And next, if $a > b > 0$, we have: $c = \sqrt{a^2 - b^2}$, and if $b > a > 0$, we have: $c = \sqrt{b^2 - a^2}$.

So either way, if $c = 0$, we get: $a = b$, that is, both the major and minor radii are the same.

Therefore, if $e = 0$, it would be not an ellipse but a circle with a radius of a (or b).

What if this time, the eccentricity $e = 1$?

If first, $a > b > 0$, we have: $e = c/a$.   So if $e = 1$, we would get: $c = a$.

And if $a > b > 0$, we have: $c = \sqrt{a^2 - b^2}$.   So if $c = a$, we would get: $b = 0$.

And if next, $b > a > 0$, we have: $e = c/b$.   So if $e = 1$, we would get: $c = b$.

And if $b > a > 0$, we have: $c = \sqrt{b^2 - a^2}$.   So if $c = b$, we would get: $a = 0$.

So either way, if the eccentricity $e = 1$, the minor axis would be 0.

And if the minor axis were 0, it would be not an ellipse but a line segment, which is in fact, the major axis, and the foci would be at both ends of the line segment.

So it would be a line segment with the length of the major axis if the eccentricity $e = 1$.

How come though, if $e = 1$, the foci would be at both ends of the major axis?

Let's take another look at the eccentricity, but this time, a bit differently.

If an ellipse is horizontal, that is, if $a > b > 0$, we have: $e = c/a$, and $c = \sqrt{a^2 - b^2}$.

Then, we can put the eccentricity e this way, too: $e = \dfrac{\sqrt{a^2 - b^2}}{a} = \sqrt{\dfrac{a^2 - b^2}{a^2}} = \sqrt{1 - \dfrac{b^2}{a^2}}$.

Then, as b changes and approaches a, the eccentricity e changes and approaches 0. And since: $c = \sqrt{a^2 - b^2}$, the focal distance c approaches 0. Thus, the foci approach the center. So as b changes the way above, the ellipse changes and approaches a circle of radius a.

Next, as b changes and approaches 0, the eccentricity e changes and approaches 1. And since: $c = \sqrt{a^2 - b^2}$, the focal distance c approaches a. Thus, the foci approach both ends of the main axis.

So as b changes the way above, the ellipse changes and approaches a line segment of length $2a$, which is the major axis, and the foci approach both ends of the line segment.

And the same is true for an ellipse vertical, too, that is, the case where: $b > a > 0$.

If an ellipse is vertical, that is, if $b > a > 0$, we have: $e = c/b$, and $c = \sqrt{b^2 - a^2}$.

Then, we can put the eccentricity e this way, too: $e = \dfrac{\sqrt{b^2 - a^2}}{b} = \sqrt{\dfrac{b^2 - a^2}{b^2}} = \sqrt{1 - \dfrac{a^2}{b^2}}$.

Then, as a changes and approaches b, the eccentricity e changes and approaches 0. And since: $c = \sqrt{b^2 - a^2}$, the focal distance c approaches 0. Thus, the foci approach the center. So as a changes the way above, the ellipse changes and approaches a circle of radius b.

Next, as a changes and approaches 0, the eccentricity e changes and approaches 1. And since: $c = \sqrt{b^2 - a^2}$, the focal distance c approaches b. Thus, the foci approach both ends of the main axis.

So as a changes the way above, the ellipse changes and approaches a line segment of length $2b$, which is the major axis, and the foci approach both ends of the line segment.

• And next, as in the cases of parabolas, ellipses have **directrices**, too.

Unlike a parabola though, an ellipse has not one but two directrices.

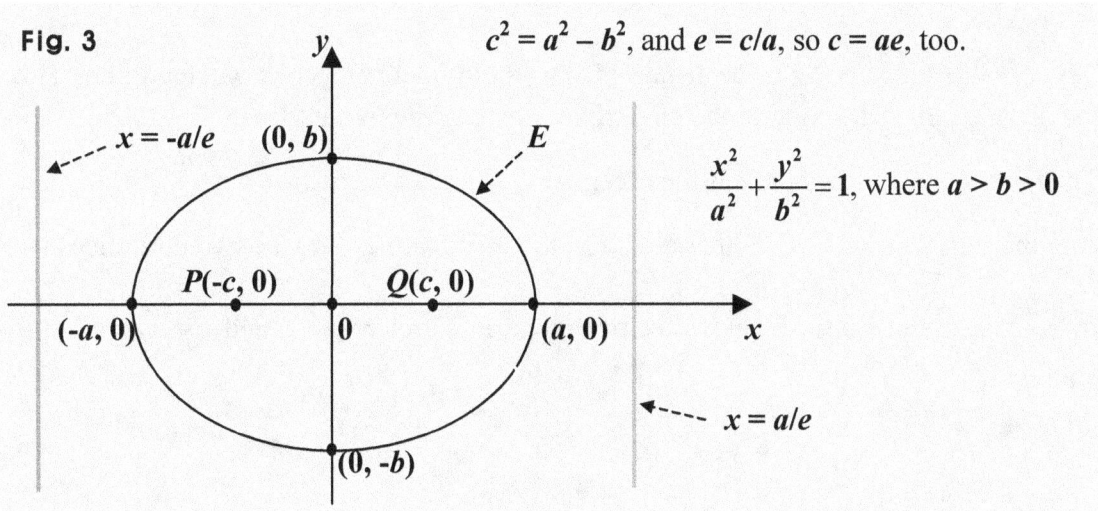

Fig. 3

$c^2 = a^2 - b^2$, and $e = c/a$, so $c = ae$, too.

$\dfrac{x^2}{a^2} + \dfrac{y^2}{b^2} = 1$, where $a > b > 0$

So as we can see in the figure above, the two vertical lines in gray are the two directrices of the ellipse E, one is on the left of the ellipse, and the other is on the right of the ellipse.

And if an ellipse is horizontal, a directrix is a vertical line, that is, a line perpendicular to the x-axis. And the focus on the left is said to correspond to the directrix on the left, and the focus on the right is said to correspond to the directrix on the right.

So in the figure above, the vertical line $x = -a/e$ is the directrix corresponding to the focus P, and the other line $x = a/e$ is the directrix corresponding to the focus Q.

How come though, the two directrices of E are the two lines $x = -a/e$, and $x = a/e$?

To begin with, in a parabola, the distance from a point to the focus is the same as the distance from the point to the directrix.

In an ellipse however, the distance from a point to a focus is less than the distance from the point to the directrix corresponding to the focus. It's because: $0 < e < 1$ for an ellipse.

So next, we can use the fact above the way below:

Suppose first, **d** is the distance from a point in a parabola to the focus, and **D** is the distance from the point to the directrix.

Then, we get: **d = D**. In other words, we get: $\frac{d}{D} = 1$.

Suppose this time also, **d** is the distance from a point in an ellipse to one focus, and **D** is the distance from the point to the directrix corresponding to the focus.

Then, we get: **d = eD** where **e** is the eccentricity. In other words, we get: $\frac{d}{D} = e$.

And the same is true for the other directrix, too. What then, are the two directrices?

In the case of the ellipse **E** below, the two directrices are: **x = -a/e**, and **x = a/e**. Why?

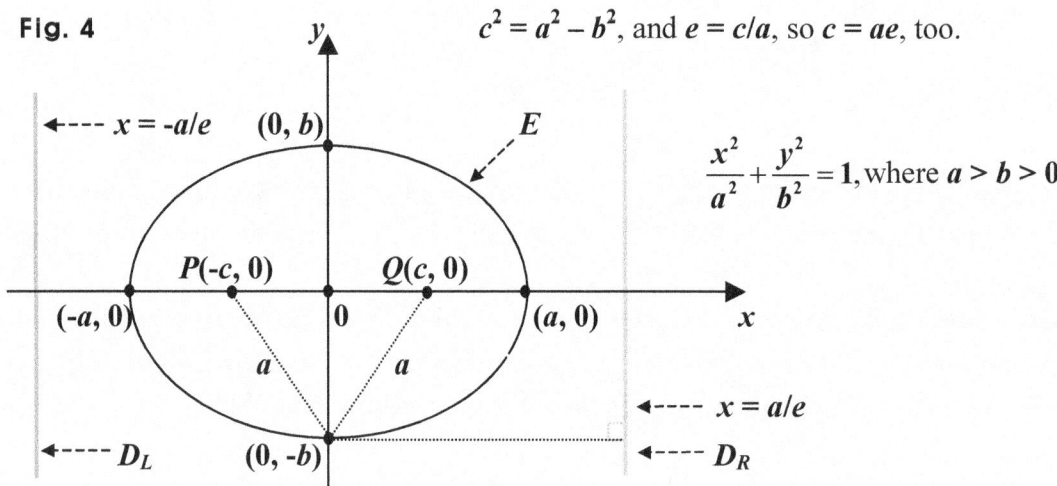

Fig. 4

$c^2 = a^2 - b^2$, and $e = c/a$, so $c = ae$, too.

$\frac{x^2}{a^2} + \frac{y^2}{b^2} = 1$, where $a > b > 0$

Let's now check to see if the directrices are: **x = -a/e**, and **x = a/e**.

Suppose first, **x = m** is the equation of the directrix **D_R** shown in the figure above.

Then, in the figure above, we can see that **m** is the distance from **(0, -b)** (which is a point in **E**) to the directrix **D_R**, and **a** is the distance from **(0, -b)** to the focus **Q**.

Next, we have: **d = eD** where **d** is the distance from a point to a focus, and **D** is the distance from the point to the directrix corresponding to the focus. So **d = a**, and **D = m**.

Thus, we get: $d = eD \Rightarrow a = em \Rightarrow m = a/e$. So the directrix **$D_R$** is the line **x = a/e**.

Let's use this time, another point $(a, 0)$ to check to see if it is the case.

Then, first, taking the distance from $(a, 0)$ to the focus Q, we get: $a - ae$, since: $c = ae$.

Next, taking the distance from $(a, 0)$ to the directrix $x = m$, that is, D_R, we get: $m - a$.

And next, we have: $d = eD$ where d is the distance from a point to a focus, and D is the distance from the point to the directrix corresponding to the focus.

So we get: $d = a - ae$, and $D = m - a$.   Thus, we get:

$d = eD \Rightarrow a - ae = e(m - a) = em - ae \Rightarrow a - ae = em - ae \Rightarrow a = em \Rightarrow m = a/e$.

So the directrix D_R is the line $x = a/e$.

And the same is true for the other directrix D_L, too, because the ellipse is symmetric about the y-axis. We know however, D_L is on the left of the origin.

So the directrix D_L is the line $x = -a/e$.

And let's now find for instance, the directrices of this ellipse: $\dfrac{x^2}{4^2} + \dfrac{y^2}{2^2} = 1$.

We have: $x = \pm a/e$, $e = c/a$, and $c = \sqrt{a^2 - b^2}$. So first, we can get: $x = \pm a^2/c$. And we have: $a = 4$, and $b = 2$.

So finding first, the focal distance, we get: $c = \sqrt{a^2 - b^2} = \sqrt{4^2 - 2^2} = \sqrt{12} = 2\sqrt{3}$.

Thus, next, we can get: $a^2/c = \dfrac{16}{2\sqrt{3}} = \dfrac{8}{\sqrt{3}} = \dfrac{8\sqrt{3}}{3}$.

So the directrices are: $x = \pm \dfrac{8\sqrt{3}}{3} \approx \pm 4.6188$.

What then, about the directrices of this ellipse: $\dfrac{(x-u)^2}{a^2} + \dfrac{(y-v)^2}{b^2} = 1$, where $a > b > 0$?

We know that the ellipse **E** is: $\dfrac{x^2}{a^2}+\dfrac{y^2}{b^2}=1$, where $a > b > 0$.

So assuming the ellipse in question is **G**, we can get **G** translating the ellipse **E** in the amount of **u** in the direction of the **x**-axis, and in the amount of **v** along the **y**-axis.

And we know that the center of the ellipse **G** is **(u, v)**.
So in short, moving the center of the ellipse **E** to the point **(u, v)**, we can get **G**.

Then, the directrices of **E** will be translated exactly the way the ellipse **E** gets translated.

And we know that the directrices of **E** are: $x = \pm a/e$.

So the directrices of **G** will be as follows: $x = \pm a/e + u$.

That is, the two directrices are: $x = -a/e + u$, and $x = a/e + u$.

Fig. 5

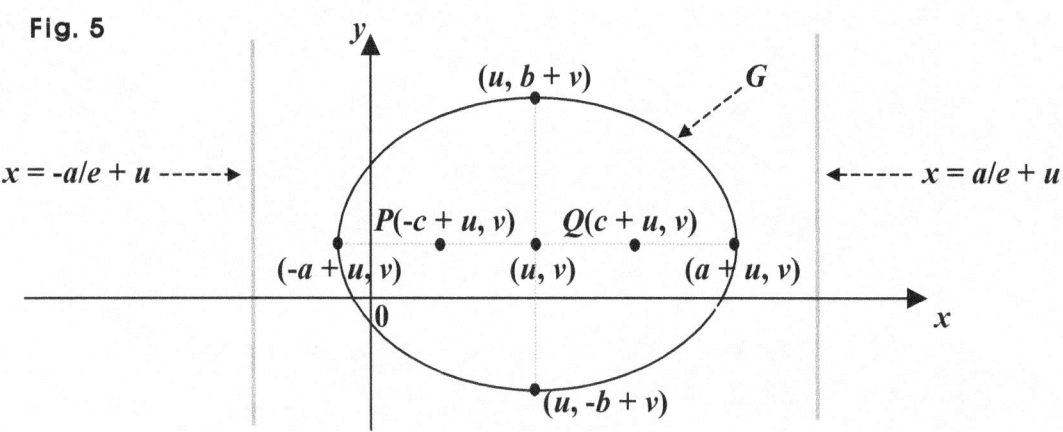

Let's now, for instance, find the directrices of this ellipse: $\dfrac{(x-2)^2}{4^2}+\dfrac{(y-1)^2}{2^2}=1$.

First, we've already found the directrices of $\dfrac{x^2}{2^2}+\dfrac{y^2}{4^2}=1$, and they are: $x = \pm\dfrac{8\sqrt{3}}{3}$.

So next, translating the directrices above by 2 along the **x**-axis, and by 1 along the **y**-axis, we get the directrices of the ellipse we want to find the directrices of.

We know however, translating a vertical line along the *y*-axis, we see no change in the line. So assuming (2, 1) is a new center, we just add the *x*-coordinate of the new center to the directrices of the ellipse <u>centered at the origin</u>.

Then, we get: $x = \pm\dfrac{8\sqrt{3}}{3} + 2$, which are the directrices we want. And more specifically:

One is: $x = -\dfrac{8\sqrt{3}}{3} + 2 = \dfrac{6 - 8\sqrt{3}}{3} \approx -2.62$, which corresponds to the left focus.

And the other is: $x = \dfrac{8\sqrt{3}}{3} + 2 = \dfrac{6 + 8\sqrt{3}}{3} \approx 6.62$, which corresponds to the right focus.

- What if the ellipse is vertical, that is, if it is: $\dfrac{x^2}{a^2} + \dfrac{y^2}{b^2} = 1$, where <u>**b > a > 0**</u>?

Putting first, the ellipse above in a graph, we can put it the way, below:

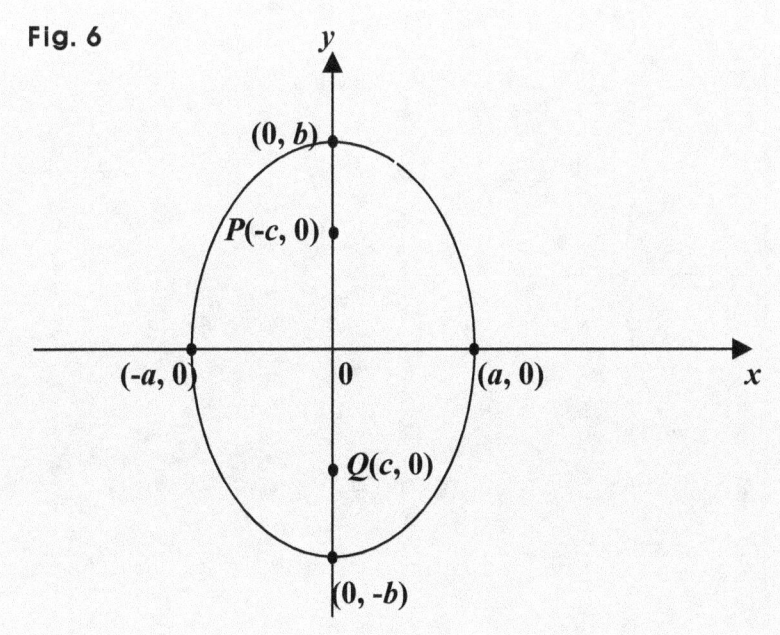

Fig. 6

To begin with, we know the fact that if in an ellipse, d is the distance from a point to a focus, and D is the distance from the point to the directrix corresponding to the focus, we get: $d = eD$, where e is the eccentricity.   In other words, we get: $\frac{d}{D} = e$.

   Next, the directrices are perpendicular to the major axis.

And we know if an ellipse is horizontal, the major axis is parallel to the x-axis.
So the directrices of an ellipse horizontal are perpendicular to the x-axis.
What then, about the directrices of an ellipse vertical?

If the ellipse is vertical, the major axis is parallel to the y-axis.
So the directrices are horizontal, that is, perpendicular to the y-axis.

And next, the directrices of E are: $x = \pm a/e$, where a is the major radius, and e is the eccentricity. What then, are the directrices of the ellipse vertical?

If the vertical ellipse is V, the directrices of V are: $y = \pm b/e$, where b is the major radius.

Fig. 7

$c^2 = b^2 - a^2$, and $e = c/b$, so $c = be$, too.

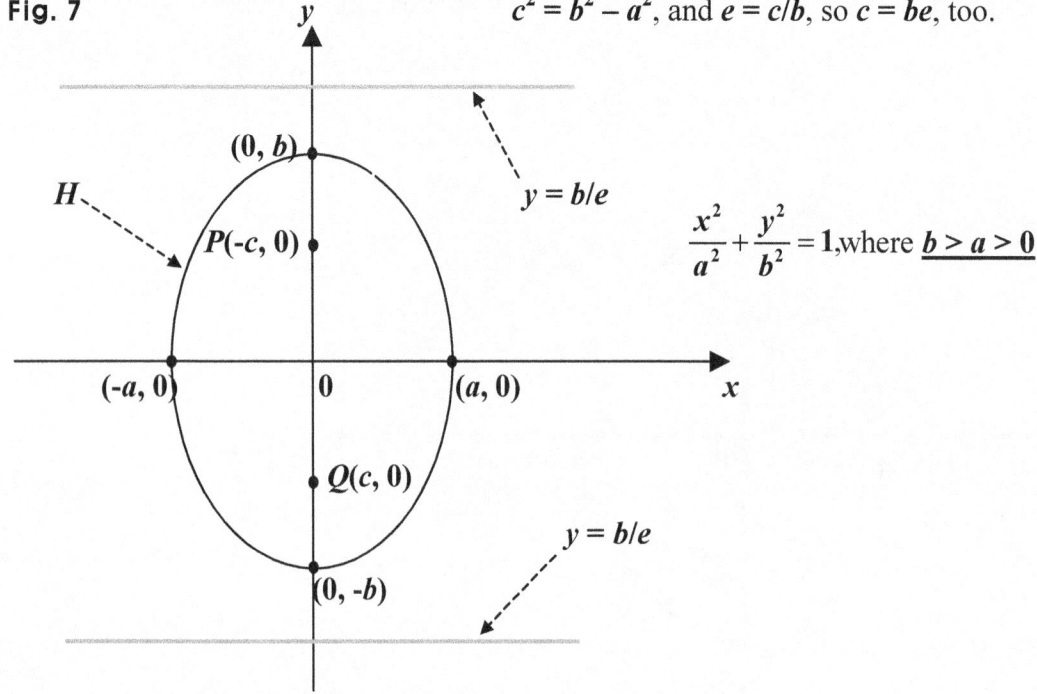

$$\frac{x^2}{a^2} + \frac{y^2}{b^2} = 1, \text{where } \underline{b > a > 0}$$

So for instance, what are the directrices of this ellipse: $\dfrac{x^2}{2^2} + \dfrac{y^2}{4^2} = 1$?

We know finding the directrices of $\dfrac{x^2}{4^2} + \dfrac{y^2}{2^2} = 1$, we get: $x = \pm\dfrac{8\sqrt{3}}{3}$.

So the directrices of $\dfrac{x^2}{2^2} + \dfrac{y^2}{4^2} = 1$ are: $y = \pm\dfrac{8\sqrt{3}}{3}$. How come?

We have: $y = \pm b/e$, $e = c/b$, and $c = \sqrt{b^2 - a^2}$. So first, we can get: $y = \pm b^2/c$.
And we have: $a = 2$, and $b = 4$.

So finding first, the focal distance, we get: $c = \sqrt{a^2 - b^2} = \sqrt{4^2 - 2^2} = \sqrt{12} = 2\sqrt{3}$.

Thus, the directrices are: $y = \pm b^2/c = \pm\dfrac{16}{2\sqrt{3}} = \pm\dfrac{8}{\sqrt{3}} = \pm\dfrac{8\sqrt{3}}{3}$. That is, $y = \pm\dfrac{8\sqrt{3}}{3}$.

• What then, about the directrices of this ellipse: $\dfrac{(x-u)^2}{a^2} + \dfrac{(y-v)^2}{b^2} = 1$, where $\underline{b > a > 0}$?

Assuming the ellipse above is J, we can get J translating the ellipse V in the amount of u in the direction of the x-axis, and in the amount of v in the direction of the y-axis.

And we know that the center of the ellipse J is (u, v).
So in short, moving the center of the ellipse V to the point (u, v), we can get J.

Then, the directrices of V will be translated exactly the way the ellipse V gets translated.

And we know that the directrices of V are: $y = \pm b/e$.

So the directrices of J will be as follows: $y = \pm b/e + v$.

That is, the two directrices are: $y = -b/e + v$, and $y = b/e + v$.

Let's now, for instance, find the directrices of this ellipse: $\dfrac{(x-3)^2}{2^2} + \dfrac{(y+5)^2}{4^2} = 1$.

First, we've already found the directrices of $\dfrac{x^2}{2^2} + \dfrac{y^2}{4^2} = 1$, and they are: $y = \pm \dfrac{8\sqrt{3}}{3}$.

So next, translating those directrices by 3 along the *x*-axis, and by -5 along the *y*-axis, we get the directrices of the ellipse we want to find the directrices of.

We know however, the directrices above are horizontal, and translating horizontal lines along the *x*-axis, we see no change.

So assuming (3, -5) is a new center, we just add the *y*-coordinate of the new center to the directrices above.   Note however, we can do so, because the ellipse getting translated is centered at the origin.

Then, we get: $y = \pm \dfrac{8\sqrt{3}}{3} - 5$, which are the directrices we want. And more specifically:

One is: $y = -\dfrac{8\sqrt{3}}{3} - 5$, which corresponds to the lower focus.

And the other is: $y = \dfrac{8\sqrt{3}}{3} - 5$, which corresponds to the upper focus.

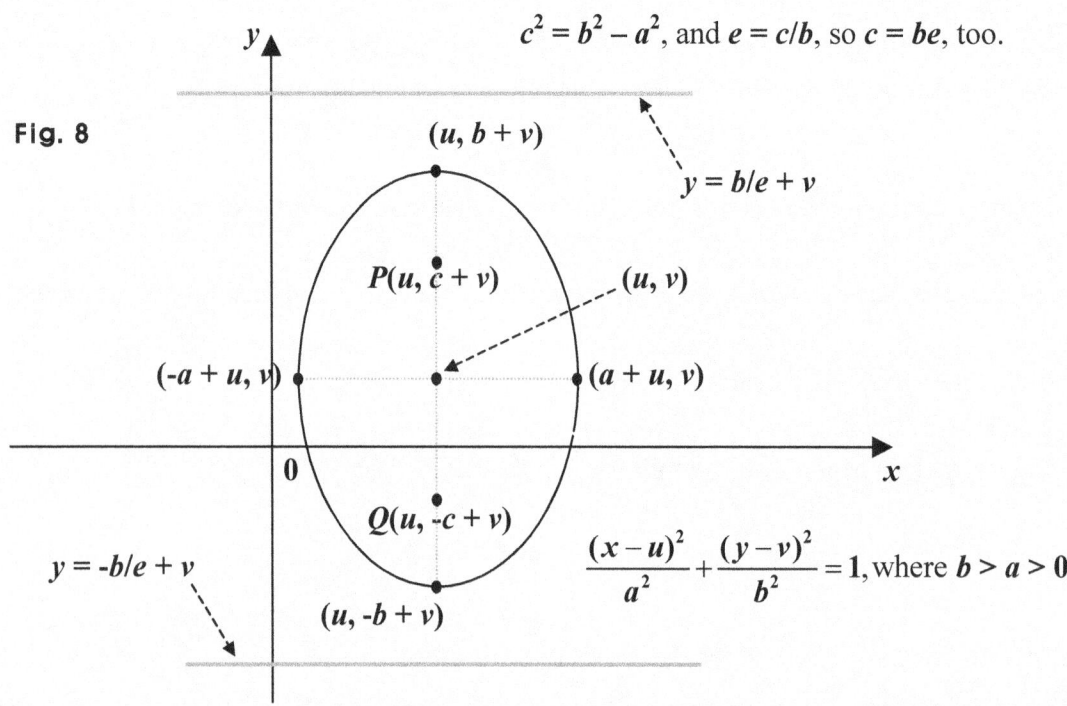

Fig. 8

$c^2 = b^2 - a^2$, and $e = c/b$, so $c = be$, too.

$(u, b + v)$

$y = b/e + v$

$P(u, c + v)$

(u, v)

$(-a + u, v)$

$(a + u, v)$

$Q(u, -c + v)$

$\dfrac{(x - u)^2}{a^2} + \dfrac{(y - v)^2}{b^2} = 1$, where $b > a > 0$

$y = -b/e + v$

$(u, -b + v)$

4. Summary on Ellipses

Now, summing up, we can put together the ideas on ellipses the way below:

To begin with, putting an ellipse in a standard equation, we can get:

$$\frac{x^2}{a^2} + \frac{y^2}{b^2} = 1, \text{ where } a > b > 0.$$ What ellipse then, is it?

If **E** is the ellipse, **E** is <u>centered at **(0, 0)**</u>, the origin, and since $a > b$, it is <u>horizontal</u>, so the major axis is parallel to the **x**-axis, and is **2a**, and the minor axis is parallel to the **y**-axis, and is **2b**.   And assuming **c** is the <u>focal distance</u>, we get: $c^2 = a^2 - b^2$.

Fig. 0

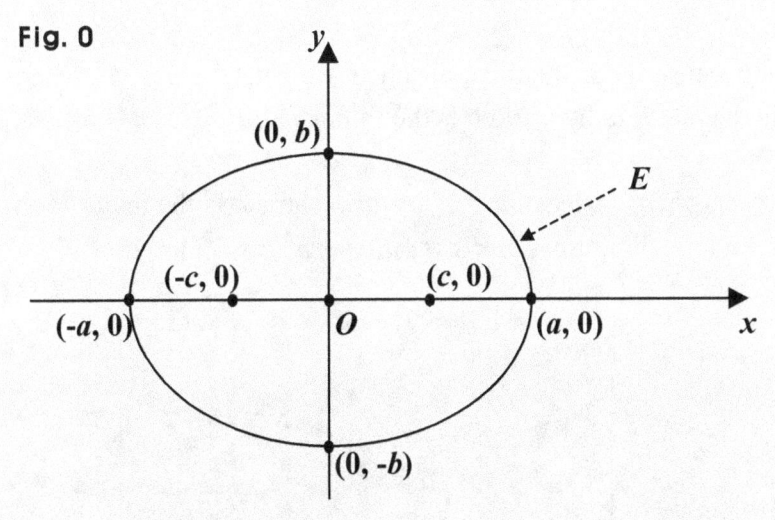

Next, since the major radius is **a**, assuming **e** is the <u>eccentricity</u>, we get: $e = c/a$, where **c** is the focal distance, of course.   And we have: $0 < e < 1$, since $0 < c < a$.

Next, since the ellipse E is horizontal, the center and foci share the same y-coordinate. And the center is the midpoint between the foci. So since the center is **(0, 0)**, and the focal distance is c, the two <u>foci</u> are **(-c, 0)** and **(c, 0)**.

And also, since E is horizontal, the center and vertices share the same y-coordinate, too. And the center is the midpoint between the vertices. The distance from the center to a vertex is the major radius. So since the major radius is a, and the center is **(0, 0)**, the two <u>vertices</u> are **(-a, 0)** and **(a, 0)**.

And next, since E is horizontal, the two directrices are two lines parallel to the y-axis. The distance from the center to each directrix is: a/e, where a is the major radius, and e is the eccentricity. So since the center is **(0, 0)**, the two directrices are: $x = \pm a/e$. And we know: $e = c/a$. So we can put the two directrices this way, too: $x = \pm a^2/c$.

So for instance, putting a horizontal ellipse in an equation, we can get: $\dfrac{x^2}{5^2} + \dfrac{y^2}{3^2} = 1$. What ellipse then, is it?

If K is the ellipse, K is <u>centered at (0, 0)</u>, the origin, and is <u>horizontal</u>, so the major axis is parallel to the x-axis, and is 10, and the minor axis is 6.

And the focal distance is 4, because if c is the <u>focal distance</u>, we get: $c^2 = a^2 - b^2$, where $a = 5$, and $b = 3$, where a is the major radius, and b is the minor radius.

So next, assuming e is the <u>eccentricity</u>, we get: $e = c/a = 4/5$, where a is the major radius, and c is the focal distance. And the directrices are: $x = \pm a/e = \pm a^2/c = 5^2/4 = 25/4$.

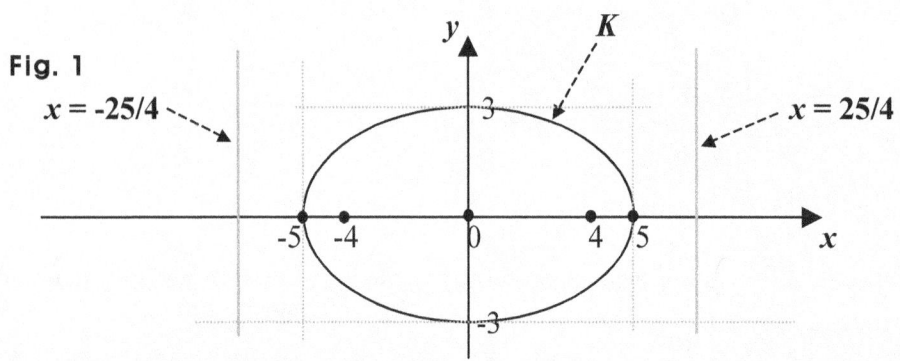

Fig. 1

Next, since the ellipse K is horizontal, the center is (0, 0), and the focal distance is 4, the two <u>foci</u> are (-4, 0) and (4, 0).   (Note that horizontal means the same y-coordinate.)

And also, since K is horizontal, the major radius, that is, the semi major axis is 5 and the center is (0, 0), the two <u>vertices</u> are (-5, 0) and (5, 0).

What then, about the ellipse as follows: $\dfrac{(x-u)^2}{a^2} + \dfrac{(y-v)^2}{b^2} = 1$, where $a > b > 0$?

If G is the ellipse, the ellipse G is <u>centered at (u, v)</u>, and since $a > b$, G is <u>horizontal</u>, so the major axis is parallel to the x-axis, and is $2a$, and the minor axis is parallel to the y-axis, and is $2b$.

And assuming c is the <u>focal distance</u>, we get: $c^2 = a^2 - b^2$, where a is the major radius, and b is the minor radius.

Next, since the major radius is a, assuming e is the <u>eccentricity</u>, we get: $e = c/a$, where c is the focal distance, of course.   And also, we have: $0 < e < 1$, too, since $0 < c < a$.

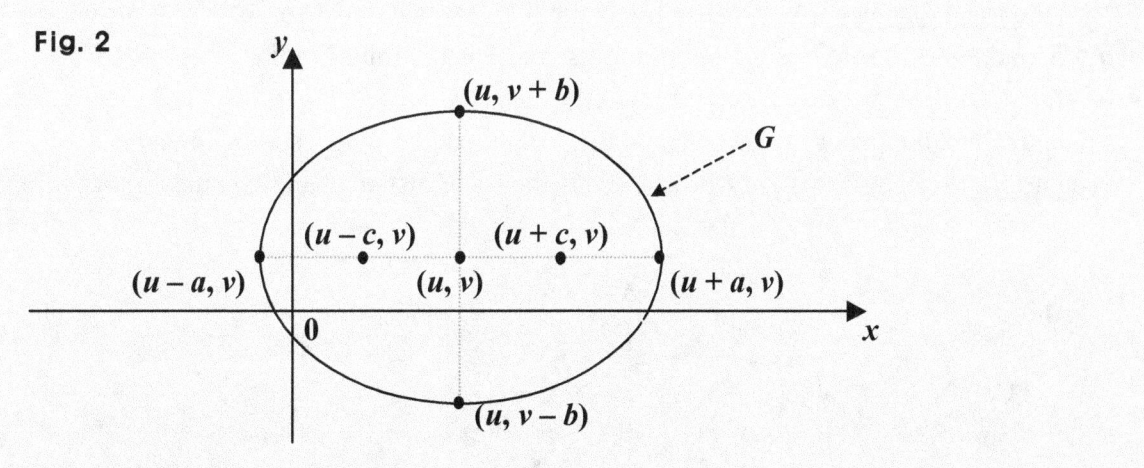

Fig. 2

And next, since G is horizontal, the two directrices are two lines parallel to the y-axis. The distance from the center to each directrix is: a/e, where a is the major radius, and e is the eccentricity. So since the center is (u, v), the two directrices are: $x = \pm a/e + u$. And we know: $e = c/a$. So we can put the two directrices this way, too: $x = \pm a^2/c + u$.

Next, since the ellipse *G* is horizontal, the center is at **(*u*, *v*)**, and the focal distance is *c*, the two <u>foci</u> are **(*u* − *c*, *v*)** and **(*u* + *c*, *v*)**.

And also, since *G* is horizontal, the center is **(*u*, *v*)**, and the major radius is *a*, the two <u>vertices</u> are **(*u* − *a*, *v*)** and **(*u* + *a*, *v*)**.

And notice that translating the ellipse *E* in the amount of *u* along the *x*-axis, and in the amount of *v* along the *y*-axis, we get the ellipse *G*.

And also, of course, translating the ellipse *G* in the amount of *-u* along the *x*-axis, and in the amount of *-v* along the *y*-axis, we get the ellipse *E*.

So for instance, putting a horizontal ellipse in an equation, we can get:

$$\frac{(x+2)^2}{5^2} + \frac{(y-1)^2}{3^2} = 1.$$ What ellipse then, is it?

If *L* is the ellipse, the ellipse *L* is <u>centered at (-2, 1)</u>, and is <u>horizontal</u>, so the major axis is parallel to the *x*-axis, and is 10, and the minor axis is 6.

And the <u>focal distance</u> is 4, because if *c* is the focal distance, we get: $c^2 = a^2 - b^2$, where *a* = 5, and *b* = 3, where *a* is the major radius, and *b* is the minor radius.

So next, since the major radius is 5, and the focal distance is 4, assuming *e* is the <u>eccentricity</u>, we get: *e* = *c*/*a* = 4/5, where *c* is the focal distance, and *a* is the major radius.

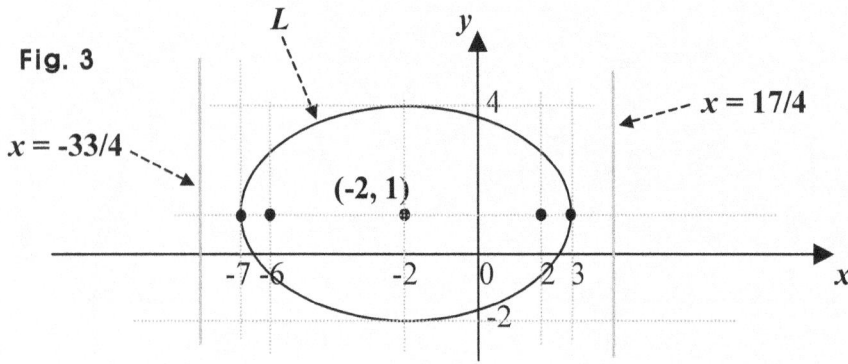

And since the center is (-2, 1), the directrices are: *x* = ±*a*/*e* − 2 = ±*a*²/*c* − 2 = ±25/4 − 2.

Next, since the ellipse L is horizontal, the focal distance is 4 and the center is (-2, 1), the two <u>foci</u> are (-6, 1) and (2, 1).

And also, since L is horizontal, the center is (-2, 1), and the major radius, that is, the semi major axis is 5, the two <u>vertices</u> are (-7, 1) and (3, 1).

And next, putting a vertical ellipse in a standard equation, we can get:

$$\frac{x^2}{a^2} + \frac{y^2}{b^2} = 1, \text{ where } b > a > 0. \qquad \text{What ellipse then, is it?}$$

If V is the ellipse, V is <u>centered at **(0, 0)**</u>, the origin, and since $b > a$, it is <u>vertical</u>, so the major axis is parallel to the y-axis, and is **2b**, and the minor axis is parallel to the x-axis, and is **2a**.   And assuming c is the <u>focal distance</u>, we get: $c^2 = b^2 - a^2$.

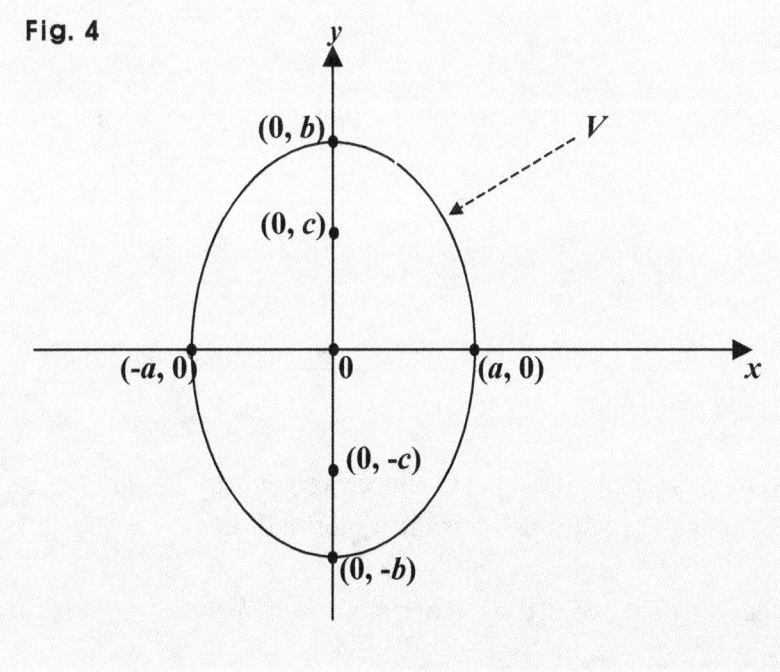

Fig. 4

Next, since the major radius is b, assuming e is the eccentricity, we get: $e = c/b$, where c is the focal distance, of course.   And also, we have: $0 < e < 1$, too, since $0 < c < b$.

Next, since the ellipse V is vertical, the center and foci share the same x-coordinate.

And the center is the midpoint between the foci. So since the center is **(0, 0)**, and the focal distance is c, the two <u>foci</u> are **(0, c)** and **(0, -c)**.

And also, since V is vertical, the center and the vertices share the same x-coordinate, too.

And the center is the midpoint between the vertices. So since the semi major axis, that is, the major radius is b, and the center is **(0, 0)**, the two <u>vertices</u> are **(0, b)** and **(0, -b)**.

And next, since V is horizontal, the two directrices are two lines parallel to the x-axis.

The distance from the center to each directrix is: b/e, where b is the major radius, and e is the eccentricity. So since the center is **(0, 0)**, the two directrices are: $x = \pm b/e$.

And we know: $e = c/b$. So we can put the two directrices this way, too: $x = \pm b^2/c$.

So for instance, putting a vertical ellipse in an equation, we can get: $\dfrac{x^2}{3^2} + \dfrac{y^2}{5^2} = 1$.
What ellipse then, is it?

If M is the ellipse, M is <u>centered at (0, 0)</u>, the origin, and is <u>vertical</u>, so the major axis is parallel to the y-axis, and is 10, and the minor axis is 6.

And the <u>focal distance</u> is 4, because if c is the focal distance, we get: $c^2 = b^2 - a^2$, where $b = 5$, and $a = 3$, where b is the major radius, and a is the minor radius.

So next, assuming e is the <u>eccentricity</u>, we get: $e = c/b = 4/5$, where b is the major radius, and c is the focal distance.

And since M is vertical, the major radius is b, and the center is **(0, 0)**, the two directrices are: $y = \pm b/e = \pm b^2/c = \pm 5^2/4 = \pm 25/4$.

Fig. 5

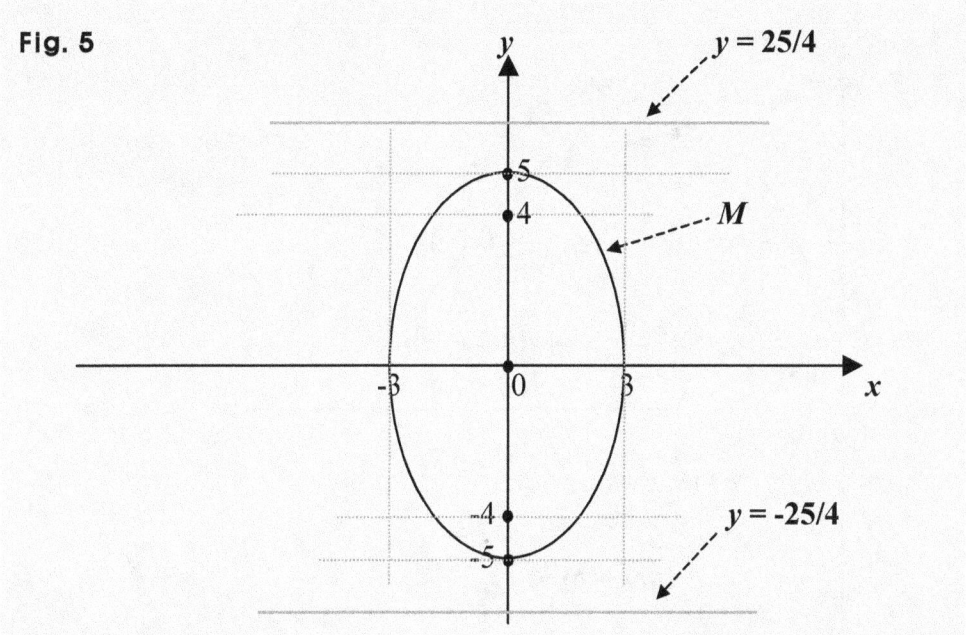

Next, since the ellipse *M* is vertical, the center is (0, 0), and the focal distance is 4, the two <u>foci</u> are (0, 4) and (0, -4).   (Note that vertical means the same *x*-coordinate.)

And also, since *M* is vertical, the semi major axis, that is, the major radius is 5 and the center is (0, 0), the two <u>vertices</u> are (0, 5) and (0, -5).

What then, about the ellipse as follows: $\dfrac{(x-u)^2}{a^2} + \dfrac{(y-v)^2}{b^2} = 1$, where $b > a > 0$?

If *J* is the ellipse, *J* is <u>centered at (*u, v*)</u>, and is <u>vertical</u>, so the major axis is parallel to the *y*-axis, and is **2b**, and the minor axis is parallel to the *x*-axis, and is **2a**.

And assuming *c* is the <u>focal distance</u>, we get: $c^2 = b^2 - a^2$.

Next, since the major radius is *b*, assuming *e* is the <u>eccentricity</u>, we get: *e* = *c/b*, where *c* is the focal distance, of course.   And also, we have: **0 < e < 1**, too, since **0 < c < b**.

And next, since *J* is vertical, the two directrices are two lines parallel to the *x*-axis. The distance from the center to each directrix is: *b/e*, where *b* is the major radius, and *e* is the eccentricity. So since the center is (*u, v*), the two directrices are: *y* = ±*b/e* + *v*.
And we know: *e* = *c/b*. So we can put the two directrices this way, too: *y* = ±*b²/c* + *v*.

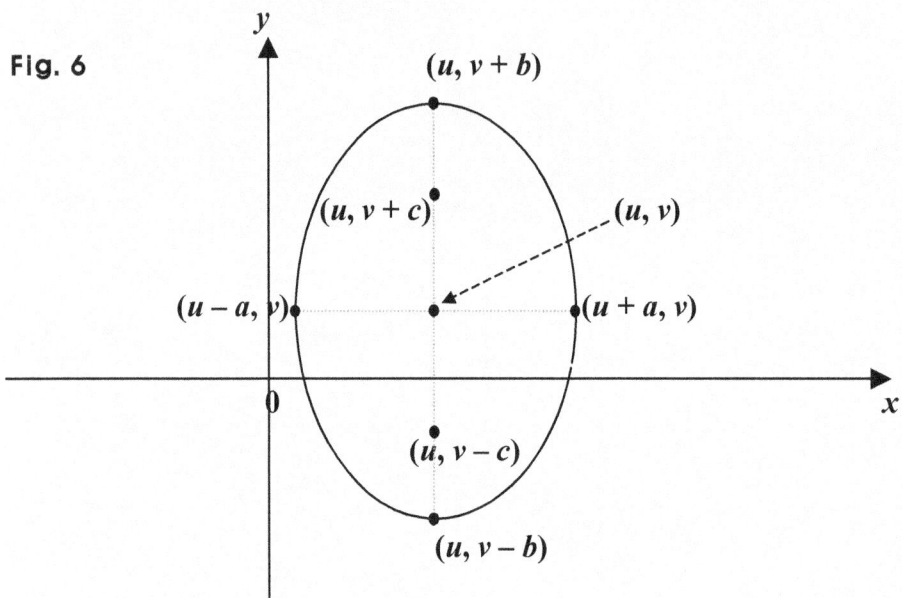

Fig. 6

Next, since *J* is vertical, the center and foci share the same *x*-coordinate. So since the center is *(u, v)*, and the focal distance is *c*, the two <u>foci</u> are *(u, v + c)* and *(u, v − c)*.

And also, since *J* is vertical, the center and the vertices share the same *x*-coordinate, too. So since the major radius, that is, the semi major axis is *b*, and the center is *(u, v)*, the two <u>vertices</u> are *(u, v + b)* and *(u, v − b)*.

So for instance, putting a vertical ellipse in an equation, we can get:

$$\frac{(x+2)^2}{3^2} + \frac{(y-1)^2}{5^2} = 1.$$ What ellipse then, is it?

If *N* is the ellipse, *N* is <u>centered at (-2, 1)</u>, and is <u>vertical</u>, so the major axis is parallel to the *y*-axis, and is 10, and the minor axis is 6.

And the <u>focal distance</u> is 4, because if *c* is the focal distance, we get: $c^2 = b^2 - a^2$, where *b* = 5, and *a* = 3, where *b* is the major radius, and *a* is the minor radius.

So next, assuming *e* is the <u>eccentricity</u>, we get: *e = c/b = 4/5*, where *b* is the major radius, and *c* is the focal distance.

And since N is vertical, the major radius is 5, and the center is **(-2, 1)**, the two directrices are: $y = \pm b/e + 1 = \pm b^2/c + 1 = \pm 5^2/4 + 1 = \pm 25/4 + 1$.

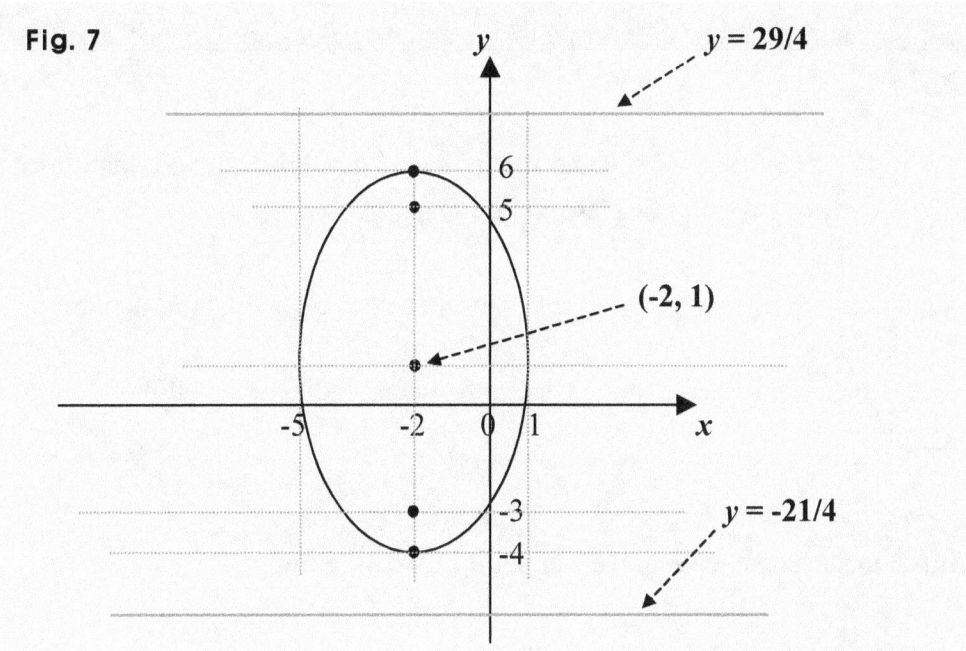

Fig. 7

Next, since the ellipse N is vertical, the center and foci share the same x-coordinate. So since the center is (-2, 1), and the focal distance is 4, the two <u>foci</u> are (-2, 5) and (-2, -3).

And also, since N is vertical, the center and the vertices share the same x-coordinate, too. So since the major radius, that is, the semi major axis is 5, and the center is (-2, 1), the two <u>vertices</u> are (-2, 6) and (-2, -4).

And we call all the equations above <u>standard equations</u> for ellipses, and the ellipses above can be said to be perpendicular, because each main axis is perpendicular to a coordinate axis.

What about an equation then, that can indicate an ellipse <u>perpendicular or not</u>?

Putting an ellipse in such an equation, we can put it the way below:

$ax^2 + by^2 + cxy + ux + vy + w = 0$, where $4ab > c^2$.

Note that *a* and *b* can be the same, and that $4ab > c^2$ can mean this, too: $ab > 0$, which is saying that *a* and *b* have the same sign. And of course, *a*, *b*, *c*, *u*, *v*, and *w* are constant.

Note however, for some values of the constants, the equation does not indicate an ellipse, even if $4ab > c^2$.

For instance, $x^2 + y^2 + xy + x + y - 1 = 0$ and $x^2 + y^2 + xy - 1 = 0$ are ellipse equations, but $x^2 + y^2 + xy + x + y + 2 = 0$ and $x^2 + y^2 + xy + 1 = 0$ are not.

That's because no real numbers for *x* and *y* can satisfy the last two equations above.

What then, about an equation that is in the general form and indicates an ellipse perpendicular only?

Putting an ellipse in such an equation, we can put it the way below:

$ax^2 + by^2 + ux + vy + w = 0$, where $a \neq b$, and $ab > 0$.

So the equation above indicates an ellipse with a main axis perpendicular to the *x*-axis. Notice that it does not have the *xy*-term, that is, *cxy* in the equation shown earlier.

Note however, for some values of the constants, the equation does not indicate an ellipse, even if $a \neq b$, and $ab > 0$.

For instance, $40x + 36y - 4x^2 - 9y^2 - 100 = 0$ and $4x^2 + 9y^2 - 1 = 0$ are ellipse equations, but $40x + 36y - 4x^2 - 9y^2 - 136 = 0$ and $4x^2 + 9y^2 + 1 = 0$ are not.

That's because:

$$4x^2 + 9y^2 + 1 = 0 \Rightarrow 4x^2 + 9y^2 = -1 \Rightarrow \frac{x^2}{3^2} + \frac{y^2}{2^2} = -1, \text{ which is not an ellipse.}$$

And in fact, no real number for *x* and *y* can satisfy the equation above.
So the equation has no curve if *x* and *y* are real.

Next, moving on to the next equation, we get:

$40x + 36y - 4x^2 - 9y^2 - 136 = 0 \Rightarrow 4x^2 + 9y^2 - 40x - 36y + 136 = 0$

$\Rightarrow 4x^2 - 40x + 9y^2 - 36y + 136 = 4(x^2 - 10x) + 9(y^2 - 4y) + 136$

$= 4(x^2 - 10x + 25 - 25) + 9(y^2 - 4y + 4 - 4) + 136$

$= 4(x^2 - 10x + 25) - 100 + 9(y^2 - 4y + 4) - 36 + 136$

$= 4(x - 5)^2 + 9(y - 2)^2 = 0 \Rightarrow \dfrac{(x - 5)^2}{3^2} + \dfrac{(y - 2)^2}{2^2} = 0,$ which is not an ellipse.

So putting more specifically a perpendicular ellipse in a general equation, we can put it the way below:

$ax^2 + by^2 + ux + vy + w = 0$, where $a \neq b$, $a > 0$, $b > 0$, and $bu^2 + av^2 > 4abw$.

That's simply because, putting the equation into the sum of perfect squares, we get:

$ax^2 + by^2 + ux + vy + w = ax^2 + ux + by^2 + vy + w$

$= a(x^2 + \dfrac{u}{a}x) + b(y^2 + \dfrac{v}{b}y) + w = a(x^2 + \dfrac{u}{a}x + \dfrac{u^2}{4a^2} - \dfrac{u^2}{4a^2}) + b(y^2 + \dfrac{v}{b}y + \dfrac{v^2}{4b^2} - \dfrac{v^2}{4b^2}) + w$

$= a(x + \dfrac{u}{2a})^2 - \dfrac{u^2}{4a} + b(y + \dfrac{v}{2b})^2 - \dfrac{v^2}{4b} + w = 0$

$\Rightarrow a(x + \dfrac{u}{2a})^2 + b(y + \dfrac{v}{2b})^2 = \dfrac{u^2}{4a} + \dfrac{v^2}{4b} - w > 0.$

Meanwhile: $\dfrac{u^2}{4a} + \dfrac{v^2}{4b} - w = \dfrac{bu^2 + av^2 - 4abw}{4ab}.$

So we get: $\dfrac{u^2}{4a} + \dfrac{v^2}{4b} - w > 0 \Rightarrow \dfrac{bu^2 + av^2 - 4abw}{4ab} > 0 \Rightarrow bu^2 + av^2 - 4abw > 0,$ since $ab > 0$.

Thus, we get: $bu^2 + av^2 > 4abw$.

Note however, in this case, we need to assume $a > 0$, and $b > 0$ as well as $a \neq b$.

Examples 1 in Ellipses

Assuming in each example below, C is the center of an ellipse, F is a focus, M is the major radius, and m is the minor radius, find the ellipse, the eccentricity, and the directrices, and put in a graph the ellipse, together with the foci, vertices, and directrices.

0. $C(0, 0)$, $F(3, 0)$, and $M = 9$.

1. $C(0, 0)$, $F(0, 3)$, and $M = 9$.

2. $C(0, 0)$, $F(3, 0)$, and $m = 9$.

3. $C(0, 0)$, $F(0, 3)$, and $m = 9$.

4. $C(0, 2)$, $F(-2, 2)$, and $M = 9$.

Suggestions or Solutions
To the Problem in the Example 0

Assuming $C(0, 0)$ is the center of an ellipse, $F(3, 0)$ is a focus, and 9 is the major radius, find the ellipse, and its elements, and put them all in a graph.

To begin with, the ellipse is horizontal, the other focus is (-3, 0), and assuming c is the focal distance, we get: $c = 3$.

So next, assuming a is the major radius, and b is the minor radius, we get:

$a = 9$, and $c^2 = a^2 - b^2 \Rightarrow 3^2 = 9^2 - b^2 \Rightarrow b^2 = 72$.

So the major axis is **18**, the minor axis is $2\sqrt{72}$, and the ellipse is: $\dfrac{x^2}{81} + \dfrac{y^2}{72} = 1$.

And the vertices are (-9, 0) and (9, 0).

Next, assuming e is the eccentricity, we get: $e = c/a = 3/9 = 1/3$.

And next, the directrices are: $x = \pm a/e = \pm a^2/c = \pm 81/3 = \pm 27$.

If not quite sure of the idea behind the processes above, follow the steps below:

To begin with, we know that the ellipse we want to find is centered at the origin.

And the standard equation of an ellipse centered at the origin is: $\dfrac{x^2}{a^2} + \dfrac{y^2}{b^2} = 1$.

So if finding the values of a and b, we find the ellipse.   How then, can we get them?

To begin with, if the ellipse is horizontal, we get: $a > b > 0$.
Then, we call a the major radius, and call b the minor radius.

If it is vertical however, we get: $b > a > 0$.
Then, we call a the minor radius, and call b the major radius.

Next, the center is (0, 0), and one of the foci is (3, 0).
So we can notice that <u>the center and the foci share the same y-coordinate</u>, which is 0.

We can see thus, the ellipse is <u>horizontal</u>.   So first, assuming the other focus is $(p, 0)$, since the center is (0, 0), and is the midpoint between the foci, we get: $0 = (p + 3)/2$.

The other focus is thus, (-3, 0). And next, assuming c is the focal distance, we get: $c = 3$, because the focal distance is the distance from the center to a focus.

Next, we can say that a is the major radius, and thus, is 9. What then, about b?

We have: $c^2 = a^2 - b^2$, where a is the major radius, and b is the minor radius.

Thus, we get: $c^2 = a^2 - b^2 \Rightarrow 3^2 = 9^2 - b^2 \Rightarrow b^2 = 81 - 9 = 72$.

So the ellipse is: $\dfrac{x^2}{81} + \dfrac{y^2}{72} = 1$, which is often put this way, of course: $\dfrac{x^2}{9^2} + \dfrac{y^2}{(\sqrt{72})^2} = 1$.

Next, the center is the midpoint between the <u>vertices</u>, too, which are the <u>endpoints</u> of the <u>major axis</u>, which is <u>twice the major radius</u>, that is, **2a**, and is 18, since **a = 9**. And since the ellipse is <u>horizontal</u>, the center and vertices share the <u>same y-coordinate</u>, too, which is 0. So since the center is (0, 0), the vertices are (-9, 0) and (9, 0).

Next, the minor axis is twice the minor radius, that is, **2b**, and thus, is $2\sqrt{72}$.

Next, the eccentricity of an ellipse is a ratio, <u>the focal distance over the major radius</u>. So assuming e is the eccentricity, we get: **e = c/a = 3/9 = 1/3**.

And next, an ellipse has two lines called the directrices, and the distance from each to the center is a ratio, which is <u>the major radius over the eccentricity</u>. So since the center is (0, 0), and the ellipse is horizontal, the directrices are: $x = \pm a/e = \pm a^2/c = \pm 81/3 = \pm 27$.

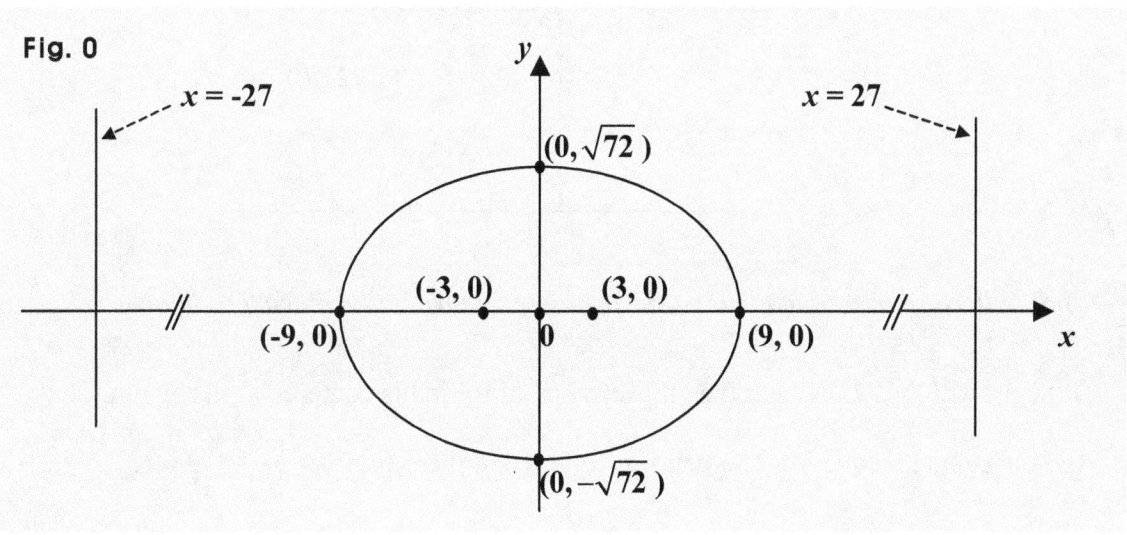

Fig. 0

Suggestions or Solutions
To the Problem in the Example 1

Assuming $C(0, 0)$ is the center of an ellipse, $F(0, 3)$ is a focus, and 9 is the major radius, find the ellipse, and its elements, and put them all in a graph.

To begin with, the ellipse is vertical, the other focus is (0, -3), and assuming c is the focal distance, we get: $c = 3$.

So next, assuming b is the major radius, and a is the minor radius, we get:

$b = 9$, and $c^2 = b^2 - a^2 \Rightarrow 3^2 = 9^2 - a^2 \Rightarrow a^2 = 72$.

So the major axis is **18**, the minor axis is $2\sqrt{72}$, and the ellipse is: $\dfrac{x^2}{72} + \dfrac{y^2}{81} = 1$.

And the vertices are (0, 9) and (0, -9).

Next, assuming e is the eccentricity, we get: $e = c/b = 1/3$.

And next, the directrices are: $y = \pm b/e = \pm b^2/c = \pm 27$.

Fig. 0

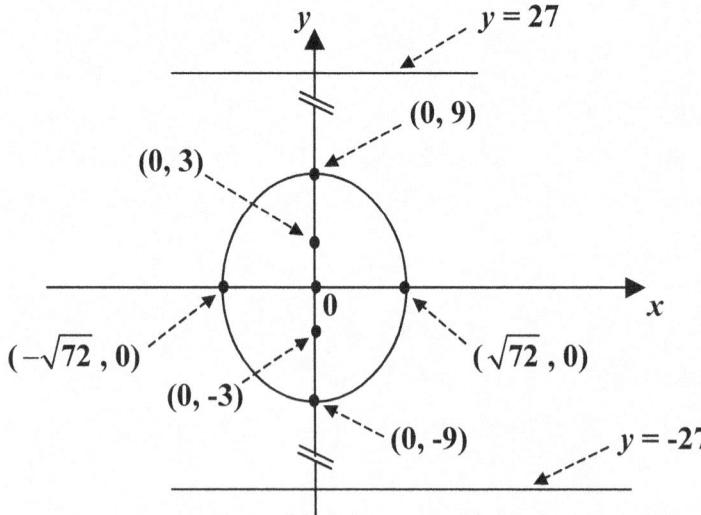

If not quite sure of the idea behind the processes above, follow the steps below:

To begin with, we know that the ellipse we want to find is centered at the origin.

And the standard equation of an ellipse centered at the origin is: $\dfrac{x^2}{a^2} + \dfrac{y^2}{b^2} = 1$.

So if finding the values of a and b, we find the ellipse.   How then, can we get them?

To begin with, if the ellipse is horizontal, we get: $a > b > 0$.
Then, we call a the major radius, and call b the minor radius.

If it is vertical however, we get: $b > a > 0$.
Then, we call a the minor radius, and call b the major radius.

Next, the center is $(0, 0)$, and one of the foci is $(3, 0)$.
So we can notice that the center and the foci share the same x-coordinate, which is 0.

We can see thus, the ellipse is vertical.   So first, assuming the other focus is $(0, q)$,
since the center is $(0, 0)$, and is the midpoint between the foci, we get: $0 = (q + 3)/2$.
The other focus is thus, $(0, -3)$. And next, assuming c is the focal distance, we get: $c = 3$,
because the focal distance is the distance from the center to a focus.

Next, we can say that b is the major radius, and thus, is 9. What then, about a?

We have: $c^2 = b^2 - a^2$, where b is the major radius, and a is the minor radius.

Thus, we get: $c^2 = b^2 - a^2 \Rightarrow 3^2 = 9^2 - a^2 \Rightarrow a^2 = 81 - 9 = 72$.

So the ellipse is: $\dfrac{x^2}{72} + \dfrac{y^2}{81} = 1$, which is often put this way, of course: $\dfrac{x^2}{(\sqrt{72})^2} + \dfrac{y^2}{9^2} = 1$.

Next, the center is the midpoint between the vertices, too, which are the endpoints of the major axis, which is twice the major radius, that is, $2b$, and is 18, since $b = 9$. And since the ellipse is vertical, the center and vertices share the same x-coordinate, too, which is 0. So since the center is $(0, 0)$, the vertices are $(0, -9)$ and $(0, 9)$.

Next, the minor axis is twice the minor radius, that is, $2a$, and thus, is $2\sqrt{72}$.

Next, the eccentricity of an ellipse is a ratio, the focal distance over the major radius.
So assuming e is the eccentricity, we get: $e = c/b = 3/9 = 1/3$.

And next, an ellipse has two lines called the directrices, and the distance from each to the center is a ratio, which is the major radius over the eccentricity.   So since the center is $(0, 0)$, and the ellipse is vertical, the directrices are: $y = \pm b/e = \pm b^2/c = \pm 81/3 = \pm 27$.

Suggestions or Solutions
To the Problem in the Example 2

Assuming $C(0, 0)$ is the center of an ellipse, $F(3, 0)$ is a focus, and 9 is the minor radius, find the ellipse, and its elements, and put them all in a graph.

To begin with, the ellipse is horizontal, the other focus is (-3, 0), and assuming c is the focal distance, we get: $c = 3$.

So next, assuming a is the major radius, and b is the minor radius, we get:

$b = 9$, and $c^2 = a^2 - b^2 \Rightarrow 3^2 = a^2 - 81 \Rightarrow a^2 = 9 + 81 = 90$.

So the major axis is $2\sqrt{90}$, the minor axis is **18**, and the ellipse is: $\dfrac{x^2}{90} + \dfrac{y^2}{81} = 1$.

And the vertices are $(\sqrt{90}, 0)$ and $(-\sqrt{90}, 0)$.

Next, assuming e is the eccentricity, we get: $e = c/a = \frac{\sqrt{10}}{10}$.

And next, the directrices are: $x = \pm a/e = \pm a^2/c = \pm 30$.

If not quite sure of the idea behind the processes above, follow the steps below:

To begin with, we know that the ellipse we want to find is centered at the origin.

And the standard equation of an ellipse centered at the origin is: $\dfrac{x^2}{a^2} + \dfrac{y^2}{b^2} = 1$.

So if finding the values of a and b, we find the ellipse.   How then, can we get them?

To begin with, if the ellipse is horizontal, we get: $a > b > 0$.
Then, we call a the major radius, and call b the minor radius.

If it is vertical however, we get: $b > a > 0$.
Then, we call a the minor radius, and call b the major radius.

Next, the center is (0, 0), and one of the foci is (3, 0).
So we can notice that <u>the center and the foci share the same y-coordinate</u>, which is 0.

We can see thus, the ellipse is <u>horizontal</u>.   So first, assuming the other focus is $(p, 0)$, since the center is (0, 0), and is the midpoint between the foci, we get: $0 = (p + 3)/2$.

The other focus is thus, (-3, 0). And next, assuming c is the focal distance, we get: $c = 3$, because the focal distance is the distance from the center to a focus.

Next, we can say that b is the minor radius, and thus, is 9. What then, about a?

We have: $c^2 = a^2 - b^2$, where a is the major radius, and b is the minor radius.

Thus, we get: $c^2 = a^2 - b^2 \Rightarrow 3^2 = a^2 - 9^2 \Rightarrow a^2 = 9 + 81 = 90$.

So the ellipse is: $\dfrac{x^2}{90} + \dfrac{y^2}{81} = 1$, which is often put this way, of course: $\dfrac{x^2}{(\sqrt{90})^2} + \dfrac{y^2}{9^2} = 1$.

Next, the center is the midpoint between the vertices, too, which are the endpoints of the <u>major axis</u>, which is <u>twice the major radius</u>, that is, $2a$, and is $2\sqrt{90}$, since $a = \sqrt{90}$. And since the ellipse is <u>horizontal</u>, the center and vertices share the <u>same y-coordinate</u>, too, which is 0.   So since the center is (0, 0), the vertices are $(\sqrt{90}, 0)$ and $(-\sqrt{90}, 0)$.

Next, the minor axis is twice the minor radius, that is, $2b$, and thus, is 18.

Next, the eccentricity of an ellipse is a ratio, <u>the focal distance over the major radius</u>.

So assuming e is the eccentricity, we get: $e = c/a = \frac{3}{\sqrt{90}} = \frac{3}{3\sqrt{10}} = \frac{1}{\sqrt{10}} = \frac{\sqrt{10}}{10}$.

And next, an ellipse has two lines called the directrices, and the distance from each to the center is a ratio, which is <u>the major radius over the eccentricity</u>.   So since the center is (0, 0), and the ellipse is horizontal, the directrices are: $x = \pm a/e = \pm a^2/c = \pm 90/3 = \pm 30$.

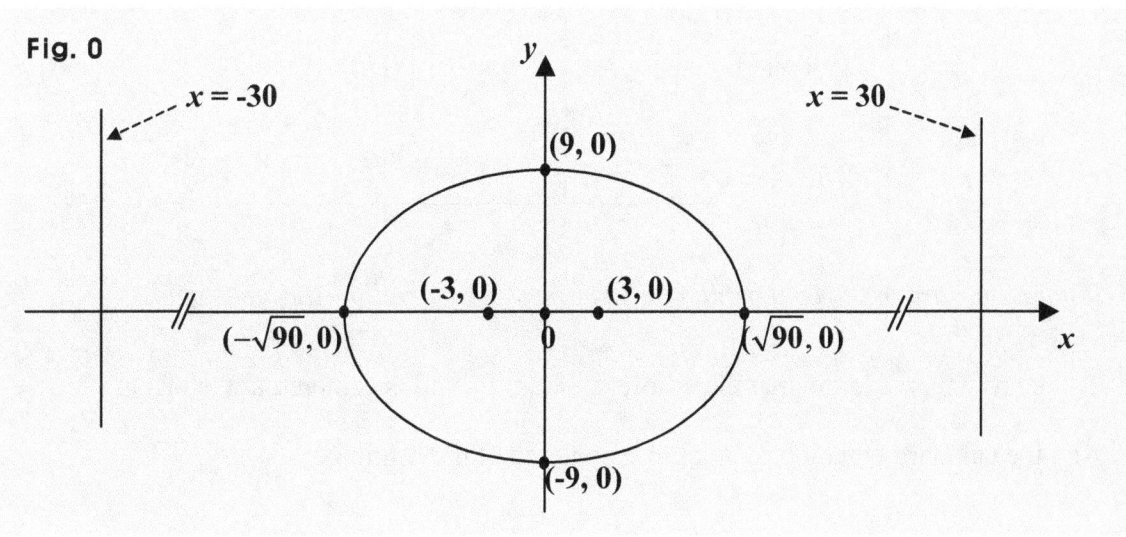

Fig. 0

Suggestions or Solutions
To the Problem in the Example 3

Assuming $C(0, 0)$ is the center of an ellipse, $F(0, 3)$ is a focus, and 9 is the minor radius, find the ellipse, and its elements, and put them all in a graph.

To begin with, the ellipse is vertical, the other focus is $(0, -3)$, and assuming c is the focal distance, we get: $c = 3$.

So next, assuming b is the major radius, and a is the minor radius, we get:

$a = 9$, and $c^2 = b^2 - a^2 \Rightarrow 3^2 = b^2 - 9^2 \Rightarrow a^2 = 90$.

So the major axis is $2\sqrt{90}$, the minor axis is **18**, and the ellipse is: $\dfrac{x^2}{81} + \dfrac{y^2}{90} = 1$.

And the vertices are $(0, \sqrt{90})$ and $(0, -\sqrt{90})$.

Next, assuming e is the eccentricity, we get: $e = c/b = \frac{\sqrt{10}}{10}$.

And next, the directrices are: $y = \pm b/e = \pm b^2/c = \pm 30$.

Fig. 0

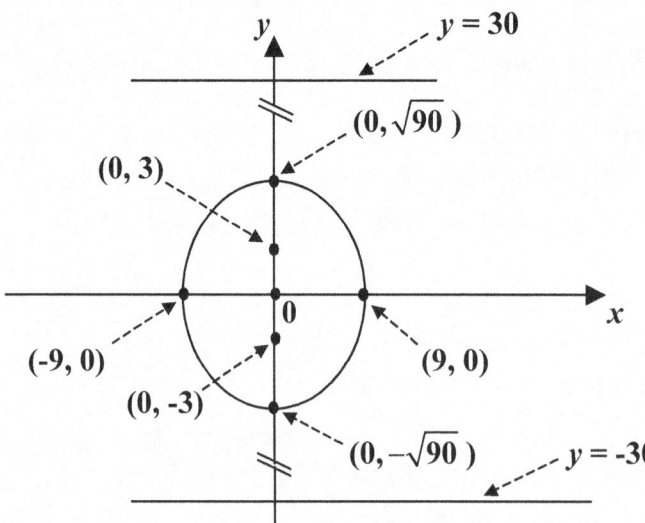

If not quite sure of the idea behind the processes above, follow the steps below:

To begin with, we know that the ellipse we want to find is centered at the origin.

And the standard equation of an ellipse centered at the origin is: $\dfrac{x^2}{a^2} + \dfrac{y^2}{b^2} = 1$.

So if finding the values of *a* and *b*, we find the ellipse.   How then, can we get them?

To begin with, if the ellipse is horizontal, we get: $a > b > 0$.
Then, we call *a* the major radius, and call *b* the minor radius.

If it is vertical however, we get: $b > a > 0$.
Then, we call *a* the minor radius, and call *b* the major radius.

Next, the center is (0, 0), and one of the foci is (3, 0).
So we can notice that the center and the foci share the same *x*-coordinate, which is 0.

We can see thus, the ellipse is vertical.   So first, assuming the other focus is (*p*, 0),
since the center is (0, 0), and is the midpoint between the foci, we get: $0 = (p + 3)/2$.
The other focus is thus, (-3, 0). And next, assuming *c* is the focal distance, we get: $c = 3$,
because the focal distance is the distance from the center to a focus.

Next, we can say that *a* is the minor radius, and thus, is 9. What then, about *b*?

We have: $c^2 = b^2 - a^2$, where *b* is the major radius, and *a* is the minor radius.

Thus, we get: $c^2 = b^2 - a^2 \Rightarrow 3^2 = b^2 - 9^2 \Rightarrow b^2 = 9 + 81 = 90$.

So the ellipse is: $\dfrac{x^2}{81} + \dfrac{y^2}{90} = 1$, which is often put this way, of course: $\dfrac{x^2}{9^2} + \dfrac{y^2}{(\sqrt{90})^2} = 1$.

Next, the center is the midpoint between the vertices, too, which are the endpoints of the major axis, which is twice the major radius, that is, **2b**, and is $2\sqrt{90}$, since $b = \sqrt{90}$. And since the ellipse is vertical, the center and vertices share the same *x*-coordinate, too, which is 0.   So since the center is (0, 0), the vertices are $(0, \sqrt{90})$ and $(0, -\sqrt{90})$.

Next, the minor axis is twice the minor radius, that is, **2a**, and thus, is 18.

Next, the eccentricity of an ellipse is a ratio, the focal distance over the major radius.

So assuming *e* is the eccentricity, we get: $e = c/b = \frac{3}{\sqrt{90}} = \frac{3}{3\sqrt{10}} = \frac{1}{\sqrt{10}} = \frac{\sqrt{10}}{10}$.

And next, an ellipse has two lines called the directrices, and the distance from each to the center is a ratio, which is the major radius over the eccentricity.   So since the center is (0, 0), and the ellipse is vertical, the directrices are: $y = \pm b/e = \pm b^2/c = \pm 90/3 = \pm 30$.

Suggestions or Solutions
To the Problem in the Example 4

Assuming $C(0, 2)$ is the center of an ellipse, $F(-2, 2)$ is a focus, and 9 is the major radius, find the ellipse, and its elements, and put them all in a graph.

To begin with, the ellipse is horizontal, the other focus is $(2, 2)$, and assuming c is the focal distance, we get: $c = 2$.

So next, assuming a is the major radius, and b is the minor radius, we get:

$a = 9$, and $c^2 = a^2 - b^2 \Rightarrow 2^2 = 9^2 - b^2 \Rightarrow b^2 = 77$.

So the major axis is **18**, the minor axis is $2\sqrt{77}$, and the ellipse is: $\dfrac{x^2}{81} + \dfrac{(y-1)^2}{77} = 1$.

And the vertices are $(-9, 2)$ and $(9, 2)$.

Next, assuming e is the eccentricity, we get: $e = c/a = 2/9$.

And next, the directrices are: $x = \pm a/e = \pm a^2/c = \pm 81/2$.

If not quite sure of the idea behind the processes above, follow the steps below:

To begin with, we know that the ellipse we want to find is centered at $(0, 2)$.

And the standard equation of an ellipse centered at $(0, 2)$ is: $\dfrac{x^2}{a^2} + \dfrac{(y-2)^2}{b^2} = 1$.

So if finding the values of a and b, we find the ellipse. How then, can we get them?

To begin with, if the ellipse is horizontal, we get: $a > b > 0$.
Then, we call a the major radius, and call b the minor radius.

If it is vertical however, we get: $b > a > 0$.
Then, we call a the minor radius, and call b the major radius.

Next, the center is $(0, 2)$, and one of the foci is $(-2, 2)$.
So we can notice that the center and the foci share the same y-coordinate, which is 2.

We can see thus, the ellipse is horizontal.   So first, assuming the other focus is $(p, 2)$, since the center is $(0, 2)$, and is the midpoint between the foci, we get: $0 = \{p + (-2)\}/2$.

The other focus is thus, (2, 2). And next, assuming c is the focal distance, we get: $c = 2$, because the focal distance is the distance from the center to a focus.

Next, we can say that a is the major radius, and thus, is 9. What then, about b?

We have: $c^2 = a^2 - b^2$, where a is the major radius, and b is the minor radius.

Thus, we get: $c^2 = a^2 - b^2 \Rightarrow 2^2 = 9^2 - b^2 \Rightarrow b^2 = 81 - 4 = 77$.

So the ellipse is: $\dfrac{x^2}{81} + \dfrac{(y-2)^2}{77} = 1$, which is often put this way: $\dfrac{x^2}{9^2} + \dfrac{(y-2)^2}{(\sqrt{77})^2} = 1$.

Next, the center is the midpoint between the vertices, too, which are the endpoints of the major axis, which is twice the major radius, that is, $2a$, and is 18, since $a = 9$. And since the ellipse is <u>horizontal</u>, the center and vertices share the <u>same y-coordinate</u>, too, which is 2.   So since the center is (0, 2), the vertices are (-9, 2) and (9, 2).

Next, the minor axis is twice the minor radius, that is, $2b$, and thus, is $2\sqrt{77}$.

Next, the eccentricity of an ellipse is a ratio, <u>the focal distance over the major radius</u>. So assuming e is the eccentricity, we get: $e = c/a = 2/9$.

And next, an ellipse has two lines called the directrices, and the distance from each to the center is a ratio, which is <u>the major radius over the eccentricity</u>.   So since the center is (0, 2), and the ellipse is horizontal, the directrices are: $x = \pm a/e = \pm a^2/c = \pm 81/2$.

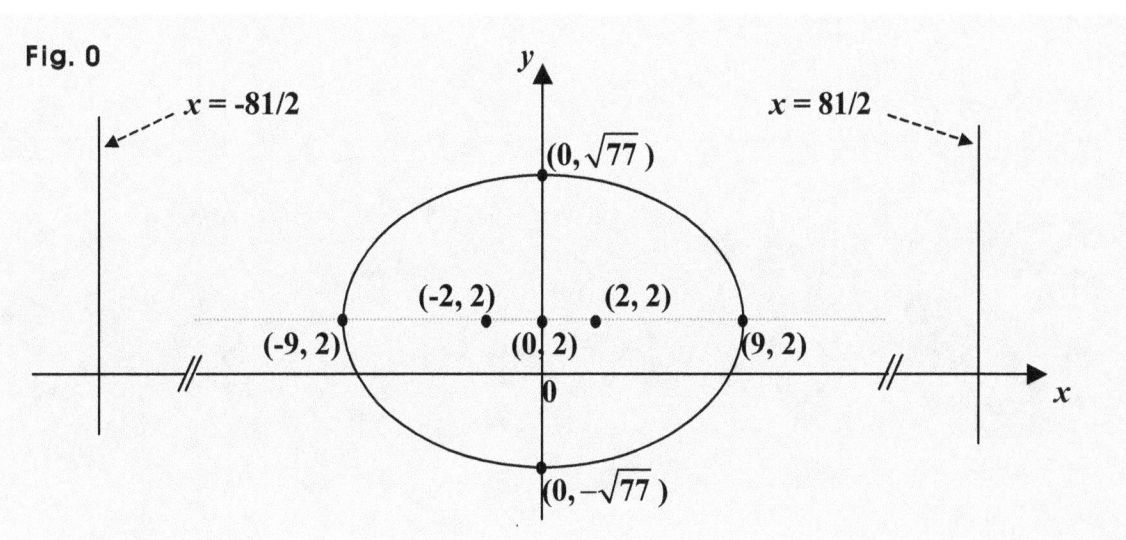

Examples 2 in Ellipses

Assuming in each example below, *C* is the center of an ellipse, *F* is a focus, *M* is the major radius, and *m* is the minor radius, find the ellipse, the eccentricity, and the directrices, and put in a graph the ellipse, together with the foci, vertices, and directrices.

0. *C*(2, 2), *F*(-2, 2), and *m* = 9.

1. *C*(3, 2), *F*(3, -3), and *M* = 9.

2. *C*(1, 2), *F*(1, 5), and *m* = 9.

3. *C*(3, 2), *F*(5, 2), and *m* = 9.

Suggestions or Solutions
To the Problem in the Example 0

Assuming $C(2, 2)$ is the center of an ellipse, $F(-2, 2)$ is a focus, and 9 is the minor radius, find the ellipse, and its elements, and put them all in a graph.

To begin with, the ellipse is horizontal, the other focus is $(6, 2)$, and assuming c is the focal distance, we get: $c = 4$.

So next, assuming a is the major radius, and b is the minor radius, we get:

$a = 9$, and $c^2 = a^2 - b^2 \Rightarrow 4^2 = 9^2 - b^2 \Rightarrow b^2 = 65$.

So the major axis is **18**, the minor axis is $2\sqrt{65}$, and the ellipse is: $\dfrac{(x-2)^2}{81} + \dfrac{(y-2)^2}{65} = 1$.

And the vertices are $(-7, 2)$ and $(11, 2)$.

Next, assuming e is the eccentricity, we get: $e = c/a = 4/9$.

And next, the directrices are: $x = \pm a/e + 2 = \pm a^2/c + 2 = \pm 81/4 + 2$.

If not quite sure of the idea behind the processes above, follow the steps below:

To begin with, we know that the ellipse we want to find is centered at $(2, 2)$.

And the standard equation of an ellipse centered at $(2, 2)$ is: $\dfrac{(x-2)^2}{a^2} + \dfrac{(y-2)^2}{b^2} = 1$.

So if finding the values of a and b, we find the ellipse. How then, can we get them?

To begin with, if the ellipse is horizontal, we get: $a > b > 0$.
Then, we call a the major radius, and call b the minor radius.

If it is vertical however, we get: $b > a > 0$.
Then, we call a the minor radius, and call b the major radius.

Next, the center is $(2, 2)$, and one of the foci is $(-2, 2)$.
So we can notice that <u>the center and the foci share the same y-coordinate</u>, which is 2.

We can see thus, the ellipse is <u>horizontal</u>. So first, assuming the other focus is $(p, 2)$, since the center is $(2, 2)$, and is the midpoint between the foci, we get: $2 = \{p + (-2)\}/2$.

The other focus is thus, (6, 2). And next, assuming c is the focal distance, we get: $c = 4$, because the focal distance is the distance from the center to a focus.

Next, we can say that a is the major radius, and thus, is 9. What then, about b?

We have: $c^2 = a^2 - b^2$, where a is the major radius, and b is the minor radius.

Thus, we get: $c^2 = a^2 - b^2 \Rightarrow 4^2 = 9^2 - b^2 \Rightarrow b^2 = 81 - 16 = 65$.

So the ellipse is: $\dfrac{(x-2)^2}{81} + \dfrac{(y-2)^2}{65} = 1$, often put this way: $\dfrac{(x-2)^2}{9^2} + \dfrac{(y-2)^2}{(\sqrt{65})^2} = 1$.

Next, the center is the midpoint between the vertices, too, which are the endpoints of the major axis, which is twice the major radius, that is, $2a$, and is 18, since $a = 9$. And since the ellipse is <u>horizontal</u>, the center and vertices share the <u>same y-coordinate</u>, too, which is 2.   So since the center is (2, 2), the vertices are (-9 + 2, 2) and (9 + 2, 2).

Next, the minor axis is twice the minor radius, that is, $2b$, and thus, is $2\sqrt{65}$.

Next, the eccentricity of an ellipse is a ratio, <u>the focal distance over the major radius</u>. So assuming e is the eccentricity, we get: $e = c/a = 4/9$.

And next, an ellipse has two lines called the directrices, and the distance from each to the center is a ratio: <u>the major radius over the eccentricity</u>. So since the center is (2, 2), and the ellipse is horizontal, the directrices are: $x = \pm a/e + 2 = \pm a^2/c + 2 = \pm 81/4 + 2$.

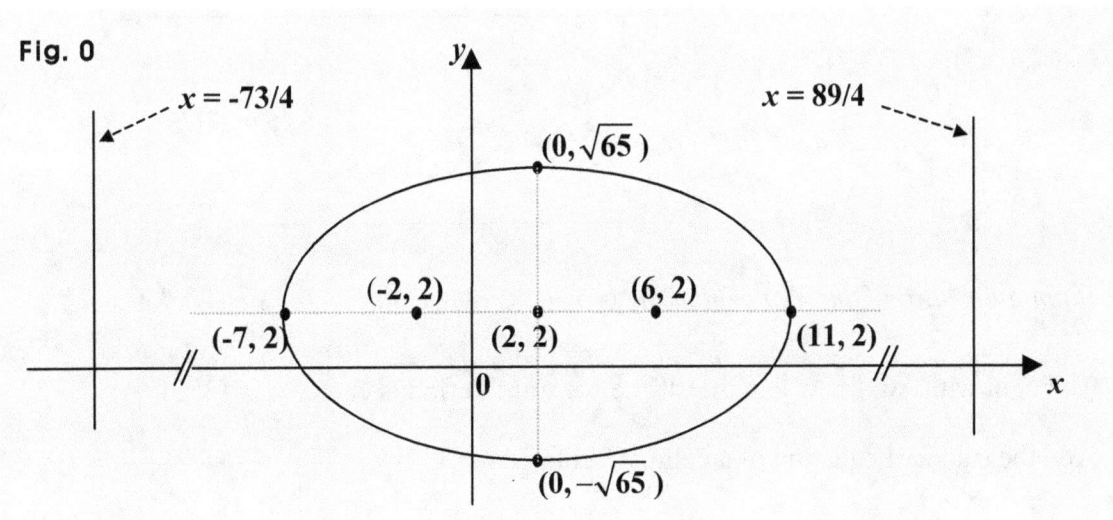

Fig. 0

Suggestions or Solutions
To the Problem in the Example 1

Assuming $C(3, 2)$ is the center of an ellipse, $F(3, -3)$ is a focus, and 9 is the major radius, find the ellipse, and its elements, and put them all in a graph.

To begin with, the ellipse is vertical, the other focus is $(3, 7)$, and assuming c is the focal distance, we get: $c = 5$.

So next, assuming b is the major radius, and a is the minor radius, we get:

$b = 9$, and $c^2 = b^2 - a^2 \Rightarrow 5^2 = 9^2 - a^2 \Rightarrow a^2 = 56$.

So the major axis is **18**, the minor axis is $2\sqrt{56}$, and the ellipse is: $\dfrac{x^2}{56} + \dfrac{y^2}{81} = 1$.

And the vertices are **(3, 11)** and **(3, -7)**.

Next, assuming e is the eccentricity, we get: $e = c/b = 5/9$.

And next, the directrices are: $y = b/e + 2 = \pm b^2/c + 2 = \pm 81/5 + 2$.

Fig. 0

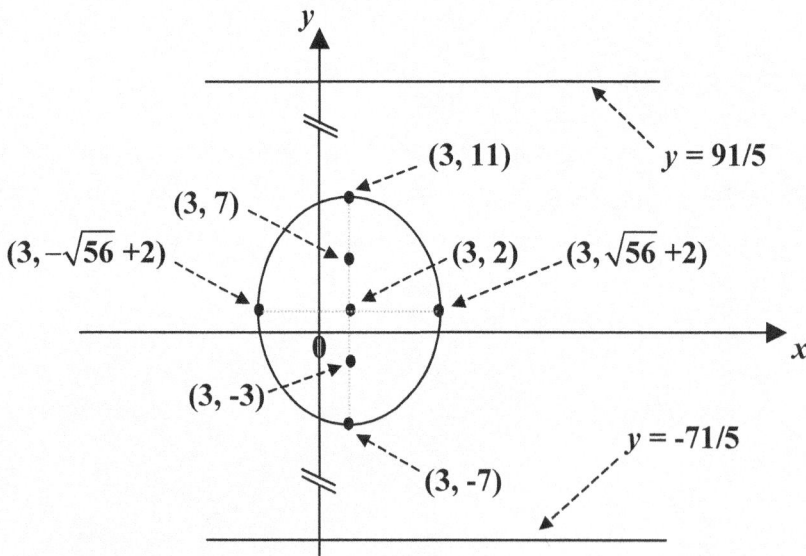

If not quite sure of the idea behind the processes above, follow the steps below:

To begin with, we know that the ellipse we want to find is centered at $(3, 2)$.

And the standard equation of an ellipse centered at $(3, 2)$ is: $\dfrac{(x-3)^2}{a^2} + \dfrac{(y-2)^2}{b^2} = 1$.

So if finding the values of *a* and *b*, we find the ellipse.   How then, can we get them?

To begin with, if the ellipse is horizontal, we get: $a > b > 0$.
Then, we call *a* the major radius, and call *b* the minor radius.

If it is vertical however, we get: $b > a > 0$.
Then, we call *a* the minor radius, and call *b* the major radius.

Next, the center is (3, 2), and one of the foci is (3, -3).
So we can notice that <u>the center and the foci share the same *x*-coordinate</u>, which is 3.

We can see thus, the ellipse is <u>vertical</u>.   So first, assuming the other focus is (3, *q*),
since the center is (3, 2), and is the midpoint between the foci, we get: $2 = \{q + (-3)\}/2$.
The other focus is thus, (3, 7). And next, assuming *c* is the focal distance, we get: $c = 5$,
because the focal distance is the distance from the center to a focus.

Next, we can say that *b* is the major radius, and thus, is 9. What then, about *a*?

We have: $c^2 = b^2 - a^2$, where *b* is the major radius, and *a* is the minor radius.

Thus, we get: $c^2 = b^2 - a^2 \Rightarrow 5^2 = 9^2 - a^2 \Rightarrow a^2 = 81 - 25 = 56$.

So the ellipse is: $\dfrac{x^2}{56} + \dfrac{y^2}{81} = 1$, which is often put this way, of course: $\dfrac{x^2}{(\sqrt{56})^2} + \dfrac{y^2}{9^2} = 1$.

Next, the center is the midpoint between the vertices, too, which are the endpoints of the major axis, which is twice the major radius, that is, **2*b***, and is 18, since $b = 9$. And since the ellipse is <u>vertical</u>, the center and vertices share the <u>same *x*-coordinate</u>, too, which is 0. So since the center is (3, 2), the vertices are **(3, 9 + 2)** and **(3, -9 + 2)**.

Next, the minor axis is twice the minor radius, that is, **2*a***, and thus, is $2\sqrt{56}$.

Next, the eccentricity of an ellipse is a ratio, <u>the focal distance over the major radius</u>.

So assuming *e* is the eccentricity, we get: $e = c/b = 5/9$.

And next, an ellipse has two lines called the directrices, and the distance from each to the center is a ratio, which is <u>the major radius over the eccentricity</u>. So since the center is (3, 2), and the ellipse is vertical, the directrices are: $y = \pm b/e + 2 = \pm b^2/c + 2 = \pm 81/5 + 2$.

Suggestions or Solutions
To the Problem in the Example 2

Assuming $C(2, 1)$ is the center of an ellipse, $F(2, 6)$ is a focus, and 9 is the minor radius, find the ellipse, and its elements, and put them all in a graph.

To begin with, the ellipse is vertical, the other focus is (2, -4), and assuming c is the focal distance, we get: $c = 5$.

So next, assuming b is the major radius, and a is the minor radius, we get:

$a = 9$, and $c^2 = b^2 - a^2 \Rightarrow 5^2 = b^2 - 9^2 \Rightarrow b^2 = 106$.

So the major axis is $2\sqrt{106}$, the minor axis is 18, and the ellipse is: $\dfrac{x^2}{81} + \dfrac{y^2}{106} = 1$.

And the vertices are $(2, \sqrt{106} + 1)$ and $(2, -\sqrt{106} + 1)$.

Next, assuming e is the eccentricity, we get: $e = c/b = \frac{5}{\sqrt{106}}$.

And next, the directrices are: $y = b/e + 1 = \pm b^2/c + 1 = \pm 106/5 + 1$.

Fig. 0

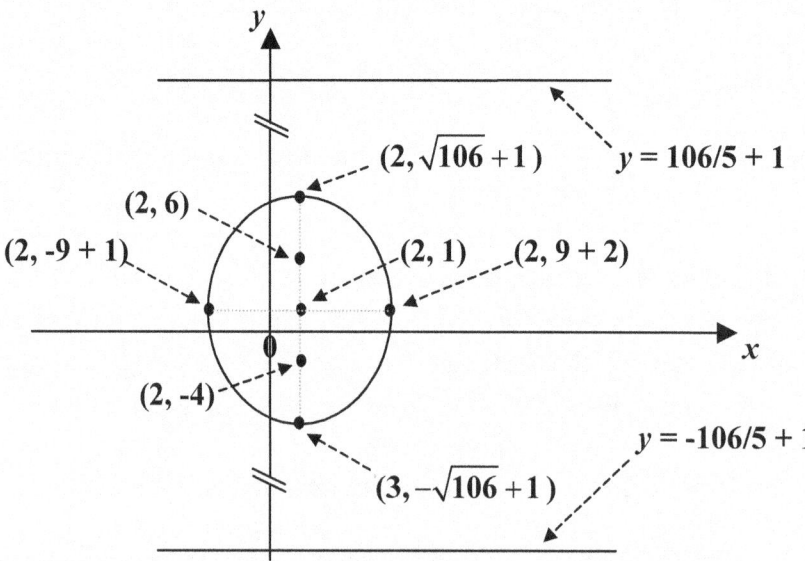

If not quite sure of the idea behind the processes above, follow the steps below:

To begin with, we know that the ellipse we want to find is centered at (2, 1).

And the standard equation of an ellipse centered at (2, 1) is: $\dfrac{(x-2)^2}{a^2} + \dfrac{(y-1)^2}{b^2} = 1.$

So if finding the values of *a* and *b*, we find the ellipse.   How then, can we get them?

To begin with, if the ellipse is horizontal, we get: $a > b > 0$.
Then, we call *a* the major radius, and call *b* the minor radius.

If it is vertical however, we get: $b > a > 0$.
Then, we call *a* the minor radius, and call *b* the major radius.

Next, the center is (2, 1), and one of the foci is (2, 6).
So we can notice that <u>the center and the foci share the same *x*-coordinate</u>, which is 2.

We can see thus, the ellipse is <u>vertical</u>.   So first, assuming the other focus is (2, *q*), since the center is (2, 1), and is the midpoint between the foci, we get: $1 = (q + 6)/2$. The other focus is thus, (2, -4). And next, assuming *c* is the focal distance, we get: $c = 5$, because the focal distance is the distance from the center to a focus.

Next, we can say that *a* is the minor radius, and thus, is 9. What then, about *b*?

We have: $c^2 = b^2 - a^2$, where *b* is the major radius, and *a* is the minor radius.

Thus, we get: $c^2 = b^2 - a^2 \Rightarrow 5^2 = b^2 - 9^2 \Rightarrow b^2 = 81 + 25 = 106$.

So the ellipse is: $\dfrac{x^2}{81} + \dfrac{y^2}{106} = 1$, which is often put this way, of course: $\dfrac{x^2}{9^2} + \dfrac{y^2}{(\sqrt{106})^2} = 1$.

Next, the center is the midpoint between the vertices, too, which are the endpoints of the major axis, which is twice the major radius, that is, $2b$, and is $2\sqrt{106}$. And since the ellipse is <u>vertical</u>, the center and vertices share the <u>same *x*-coordinate</u>, too, which is 2. So since the center is (2, 1), the vertices are $(2, \sqrt{106} + 1)$ and $(2, -\sqrt{106} + 1)$.

Next, the minor axis is twice the minor radius, that is, $2a$, and thus, is 18.

Next, the eccentricity of an ellipse is a ratio, <u>the focal distance over the major radius</u>.

So assuming *e* is the eccentricity, we get: $e = c/b = \frac{5}{\sqrt{106}}$.

And next, an ellipse has two lines called the directrices, and the distance from each to the center is a ratio: <u>the major radius over the eccentricity</u>. So since the center is (2, 1), and the ellipse is vertical, the directrices are: $y = \pm b/e + 1 = \pm b^2/c + 1 = \pm 106/5 + 1$.

Suggestions or Solutions
To the Problem in the Example 3

Assuming $C(3, 2)$ is the center of an ellipse, $F(5, 2)$ is a focus, and 9 is the minor radius, find the ellipse, and its elements, and put them all in a graph.

To begin with, the ellipse is horizontal, the other focus is $(1, 2)$, and assuming c is the focal distance, we get: $c = 2$.
So next, assuming a is the major radius, and b is the minor radius, we get:

$b = 9$, and $c^2 = a^2 - b^2 \Rightarrow 2^2 = a^2 - 9^2 \Rightarrow a^2 = 85$.

So the major axis is $2\sqrt{85}$, the minor axis is 18, and the ellipse is: $\dfrac{(x-3)^2}{85} + \dfrac{(y-2)^2}{81} = 1$.

And the vertices are $(3 - \sqrt{85}, 2)$ and $(3 + \sqrt{85}, 2)$.

Next, assuming e is the eccentricity, we get: $e = c/a = \dfrac{2}{\sqrt{85}}$.

And next, the directrices are: $x = \pm a/e + 3 = \pm a^2/c + 3 = \pm 85/2 + 3$.

If not quite sure of the idea behind the processes above, follow the steps below:

To begin with, we know that the ellipse we want to find is centered at $(3, 2)$.

And the standard equation of an ellipse centered at $(3, 2)$ is: $\dfrac{(x-3)^2}{a^2} + \dfrac{(y-2)^2}{b^2} = 1$.

So if finding the values of a and b, we find the ellipse.   How then, can we get them?

To begin with, if the ellipse is horizontal, we get: $a > b > 0$.
Then, we call a the major radius, and call b the minor radius.

If it is vertical however, we get: $b > a > 0$.
Then, we call a the minor radius, and call b the major radius.

Next, the center is $(3, 2)$, and one of the foci is $(5, 2)$.
So we can notice that the center and the foci share the same y-coordinate, which is 2.

We can see thus, the ellipse is horizontal.   So first, assuming the other focus is $(p, 2)$, since the center is $(3, 2)$, and is the midpoint between the foci, we get: $3 = (p + 5)/2$.

The other focus is thus, (1, 2). And next, assuming *c* is the focal distance, we get: *c* = **2**, because the focal distance is the distance from the center to a focus.

Next, we can say that *b* is the minor radius, and thus, is 9. What then, about *a*?

We have: $c^2 = a^2 - b^2$, where *a* is the major radius, and *b* is the minor radius.

Thus, we get: $c^2 = a^2 - b^2 \Rightarrow 2^2 = a^2 - 9^2 \Rightarrow a^2 = 81 + 4 = 85$.

So the ellipse is: $\dfrac{(x-3)^2}{85} + \dfrac{(y-2)^2}{81} = 1$, often put this way: $\dfrac{(x-3)^2}{(\sqrt{85})^2} + \dfrac{(y-2)^2}{9^2} = 1$.

Next, the center is the midpoint between the vertices, too, which are the endpoints of the major axis, which is twice the major radius, that is, **2a**, and is $2\sqrt{85}$. And since the ellipse is <u>horizontal</u>, the center and vertices share the <u>same *y*-coordinate</u>, too, which is 2. So since the center is (3, 2), the vertices are $(3 - \sqrt{85}, 2)$ and $(3 + \sqrt{85}, 2)$.

Next, the minor axis is twice the minor radius, that is, **2b**, and thus, is 18.

Next, the eccentricity of an ellipse is a ratio, <u>the focal distance over the major radius</u>.

So assuming *e* is the eccentricity, we get: $e = c/a = \frac{2}{\sqrt{85}}$.

And next, an ellipse has two lines called the directrices, and the distance from each to the center is a ratio: <u>the major radius over the eccentricity</u>. So since the center is (3, 2), and the ellipse is horizontal, the directrices are: $x = \pm a/e + 3 = \pm a^2/c + 3 = \pm 85/2 + 3$.

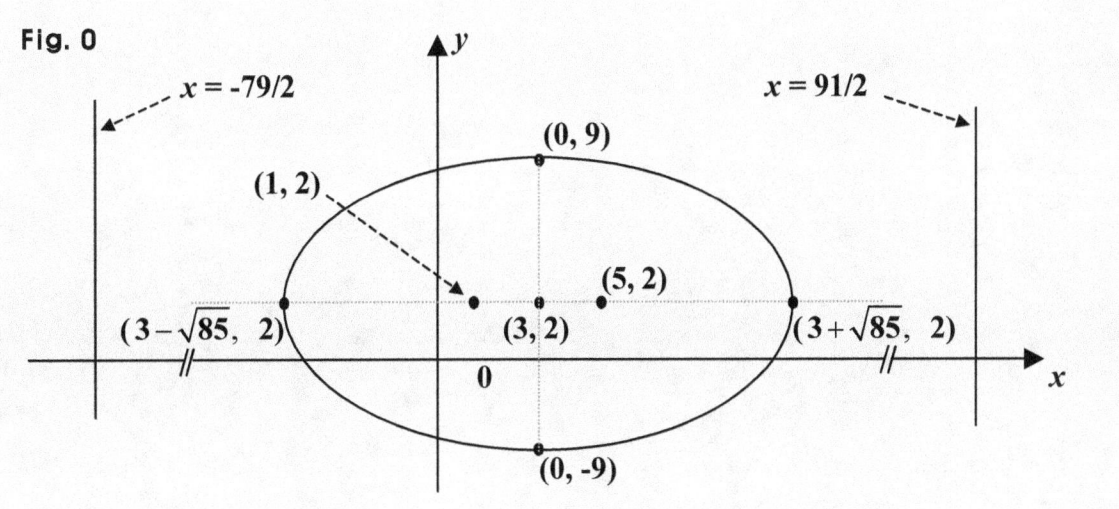

Fig. 0

Examples 3 in Ellipses

Find if each ellipse below is horizontal or vertical, and the center, foci, vertices, major axis, minor axis, eccentricity, and directrices.

0. $\dfrac{x^2}{25} + \dfrac{y^2}{16} = 1$

1. $\dfrac{x^2}{16} + \dfrac{y^2}{25} = 1$

2. $\dfrac{(x-1)^2}{25} + \dfrac{(y-2)^2}{16} = 1$

3. $25(x-1)^2 + 16(y-2)^2 = 400$

4. $\dfrac{(x-1)^2}{4} + y^2 = 1$

5. $4(x-1)^2 + (y-2)^2 = 4$

Suggestions or Solutions
To the Problem in the Example 0

Find all the elements of the ellipse as follows: $\dfrac{x^2}{25} + \dfrac{y^2}{16} = 1$.

To begin with, the ellipse is horizontal, the center is (0, 0), the major axis is 10, and the minor axis is 8.

Next, assuming c is the focal distance, a is the major radius, and b is the minor radius, we get: $a = 5$, $b = 4$, and $c^2 = a^2 - b^2 \Rightarrow c^2 = 25 - 16 = 9 \Rightarrow c = 3$.

So the focal distance is 3, and the foci are (-3, 0) and (3, 0).

And the vertices are (-5, 0) and (5, 0).

Next, assuming e is the eccentricity, we get: $e = c/a = 3/5$.

And next, the directrices are: $x = \pm a/e = \pm a^2/c = \pm 25/3$.

If not quite sure of the idea behind the processes above, follow the steps below:

To begin with, we can put the ellipse given this way: $\dfrac{x^2}{5^2} + \dfrac{y^2}{4^2} = 1$.

And the standard equation of an ellipse centered at the origin is: $\dfrac{x^2}{a^2} + \dfrac{y^2}{b^2} = 1$.
So the center of the ellipse given is (0, 0).

Next, if $a > b > 0$ in the equation above, the ellipse is horizontal, a is the major radius, and b is the minor radius.

So the ellipse given is horizontal, the major radius is 5, and the minor radius is 4.

And the major axis is twice the major radius, and thus, is 10, and the minor axis is twice the minor radius, and thus, is 8.   What then, about the focal distance?

Assuming c is the focal distance, a is the major radius, and b is the minor radius, we get: $c^2 = a^2 - b^2$.   So we get: $c^2 = a^2 - b^2 = 5^2 - 4^2 = 25 - 16 = 9 \Rightarrow c = 3$.

Next, the center, foci, and vertices are all in the major axis.
So if the ellipse is <u>horizontal</u>, the center, foci, and vertices share the <u>same y-coordinate</u>.

The center is the midpoint between the foci, and the focal distance is the distance from the center to each focus, and is c, which is 3.

And also, the center is the midpoint between the <u>vertices</u>, too, which are the <u>endpoints</u> of the <u>major axis</u>, which is twice the major radius, which is the distance from the center to each vertex, and in this case, is a, which is 5.

So since the center is (0, 0), and the ellipse is horizontal, the foci are (3, 0) and (-3, 0), and the vertices are (5, 0) and (-5, 0).

Next, the eccentricity of an ellipse is a ratio, <u>the focal distance over the major radius</u>. So assuming e is the eccentricity, we get: $e = c/a = 3/5$.

And next, an ellipse has two lines called the directrices, and the distance from each to the center is a ratio, <u>the major radius over the eccentricity</u>.   So since the center is (0, 0), and the ellipse is <u>horizontal</u>, the directrices are: $x = \pm a/e = \pm a^2/c = \pm 25/3$.

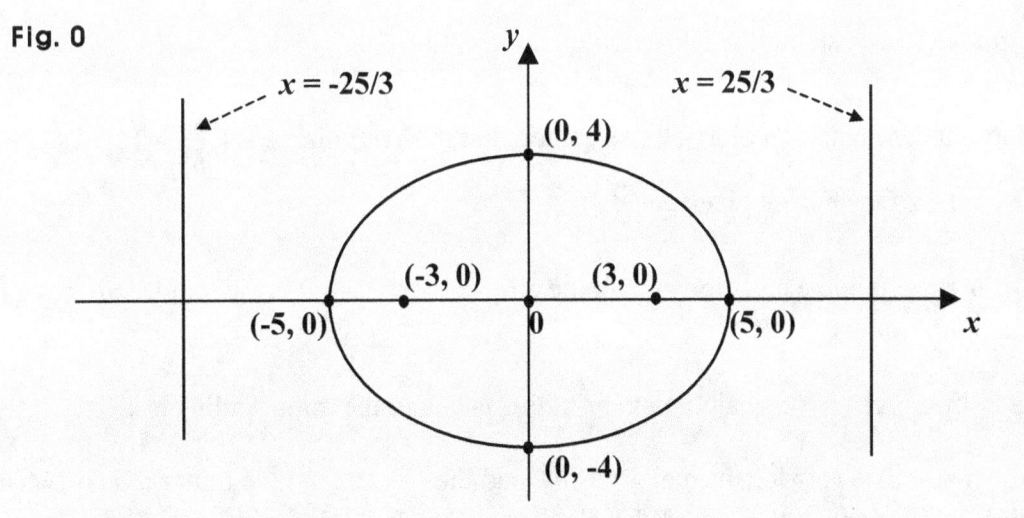

Fig. 0

The ellipse is: $\dfrac{x^2}{5^2} + \dfrac{y^2}{4^2} = 1.$

Suggestions or Solutions

To the **Problem** in the Example 1

Find all the elements of the ellipse as follows: $\dfrac{x^2}{16} + \dfrac{y^2}{25} = 1$.

To begin with, the ellipse is vertical, the center is (0, 0), the major axis is 10, and the minor axis is 8.

Next, assuming c is the focal distance, b is the major radius, and a is the minor radius, we get: $b = 5$, $a = 4$, and $c^2 = b^2 - a^2 \Rightarrow c^2 = 25 - 16 = 9 \Rightarrow c = 3$.

So the focal distance is 3, and the foci are (0, 3) and (0, -3).

And the vertices are (0, 5) and (0, -5).

Next, assuming e is the eccentricity, we get: $e = c/b = 3/5$.

And next, the directrices are: $y = \pm b/e = \pm b^2/c = \pm 25/3$.

If not quite sure of the idea behind the processes above, follow the steps below:

To begin with, we can put the ellipse given this way: $\dfrac{x^2}{4^2} + \dfrac{y^2}{5^2} = 1$.

And the standard equation of an ellipse centered at the origin is: $\dfrac{x^2}{a^2} + \dfrac{y^2}{b^2} = 1$.

So the center of the ellipse given is (0, 0).

Next, if $b > a > 0$ in the equation above, the ellipse is vertical, b is the major radius, and a is the minor radius.

So the ellipse given is <u>vertical</u>, the major radius is 5, and the minor radius is 4.

And the major axis is twice the major radius, and thus, is 10, and the minor axis is twice the minor radius, and thus, is 8. What then, about the focal distance?

Assuming c is the focal distance, b is the major radius, and a is the minor radius, we get: $c^2 = b^2 - a^2$. So we get: $c^2 = b^2 - a^2 = 5^2 - 4^2 = 25 - 16 = 9 \Rightarrow c = 3$.

Next, the center, foci, and vertices are all in the major axis.
So if the ellipse is <u>vertical</u>, the center, foci, and vertices share the <u>same *x*-coordinate</u>.

The center is the midpoint between the foci, and the focal distance is the distance from the center to each focus, and is *c*, which is 3.

And also, the center is the midpoint between the <u>vertices</u>, too, which are the <u>endpoints</u> of the <u>major axis</u>, which is twice the major radius, which is the distance from the center to each vertex, and in this case, is *b*, which is 5.

So since the center is (0, 0), and the ellipse is vertical, the foci are (0, 3) and (0, -3), and the vertices are (0, 5) and (0, -5).

Next, the eccentricity of an ellipse is a ratio, <u>the focal distance over the major radius</u>.
So assuming *e* is the eccentricity, we get: $e = c/b = \mathbf{3/5}$.

And next, an ellipse has two lines called the directrices, and the distance from each to the center is a ratio, <u>the major radius over the eccentricity</u>.   So since the center is (0, 0), and the ellipse is <u>vertical</u>, the directrices are: $y = \pm b/e = \pm b^2/c = \mathbf{\pm 25/3}$.

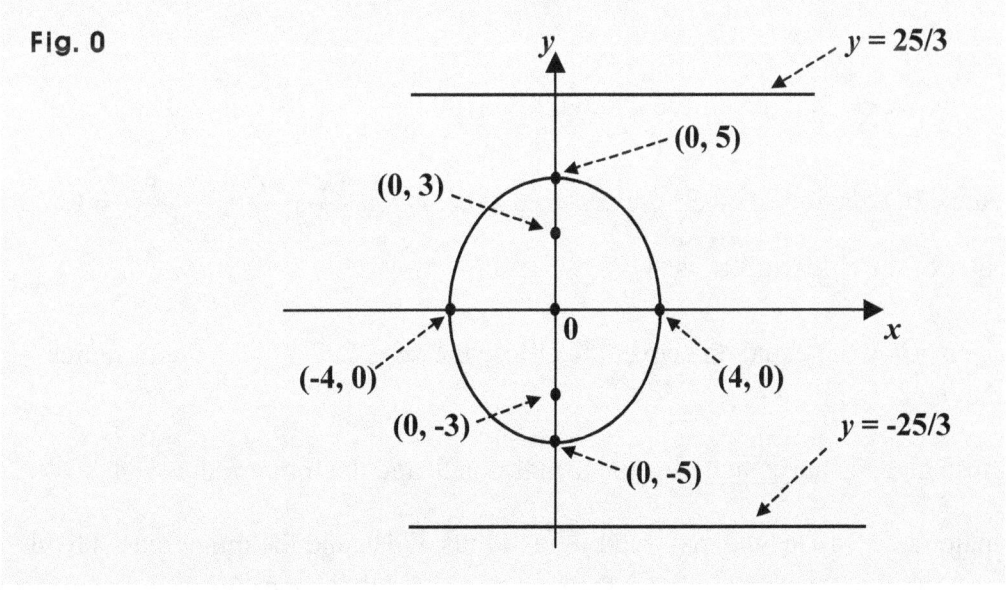

Fig. 0

The ellipse is: $\dfrac{x^2}{4^2} + \dfrac{y^2}{5^2} = 1$.

Suggestions or Solutions
To the **Problem** in the Example **2**

Find all the elements of the ellipse as follows: $\dfrac{(x-1)^2}{25}+\dfrac{(y-2)^2}{16}=1$.

To begin with, the ellipse is horizontal, the center is (1, 2), the major axis is 10, and the minor axis is 8.

Next, assuming c is the focal distance, a is the major radius, and b is the minor radius, we get: $a = 5$, $b = 4$, and $c^2 = a^2 - b^2 \Rightarrow c^2 = 25 - 16 = 9 \Rightarrow c = 3$.

So the focal distance is 3, and the foci are (-2, 2) and (4, 2).

And the vertices are (-4, 2) and (6, 2).

Next, assuming e is the eccentricity, we get: $e = c/a = 3/5$.

And next, the directrices are: $x = \pm a/e + 1 = \pm a^2/c + 1 = \pm 25/3 + 1$.

If not quite sure of the idea behind the processes above, follow the steps below:

To begin with, we can put the ellipse given this way: $\dfrac{(x-1)^2}{5^2}+\dfrac{(y-2)^2}{4^2}=1$.

And the standard equation of an ellipse centered at (u, v) is: $\dfrac{(x-u)^2}{a^2}+\dfrac{(y-v)^2}{b^2}=1$.

So the center of the ellipse given is (1, 2).

Next, if $a > b > 0$ in the equation above, the ellipse is horizontal, a is the major radius, and b is the minor radius.

So the ellipse given is <u>horizontal</u>, the major radius is 5, and the minor radius is 4.

And the major axis is twice the major radius, and thus, is 10, and the minor axis is twice the minor radius, and thus, is 8.   What then, about the focal distance?

Assuming c is the focal distance, a is the major radius, and b is the minor radius, we get: $c^2 = a^2 - b^2$.   So we get: $c^2 = a^2 - b^2 = 5^2 - 4^2 = 25 - 16 = 9 \Rightarrow c = 3$.

Next, the center, foci, and vertices are all in the major axis.
So if the ellipse is <u>horizontal</u>, the center, foci, and vertices share the <u>same y-coordinate</u>.

The center is the midpoint between the foci, and the focal distance is the distance from the center to each focus, and is c, which is 3.

And also, the center is the midpoint between the <u>vertices</u>, too, which are the <u>endpoints</u> of the <u>major axis</u>, which is twice the major radius, which is the distance from the center to each vertex, and in this case, is a, which is 5.

So since the center is (1, 2), and the ellipse is horizontal, the two foci are (3 + 1, 0 + 2) and (-3 + 1, 0 + 2), that is, **(4, 2)** and **(-2, 2)**, and by the same token, the two vertices are (5 + 1, 0 + 2) and (-5 + 1, 0 + 2), that is, **(6, 2)** and **(-4, 2)**.

Next, the eccentricity of an ellipse is a ratio, <u>the focal distance over the major radius</u>. So assuming e is the eccentricity, we get: $e = c/a = 3/5$.

And next, an ellipse has two lines called the directrices, and the distance from each to the center is a ratio, <u>the major radius over the eccentricity</u>.   So since the center is (1, 2), and the ellipse is <u>horizontal</u>, the directrices are: $x = \pm a/e + 1 = \pm a^2/c + 1 = \pm 25/3 + 1$.

Fig. 0

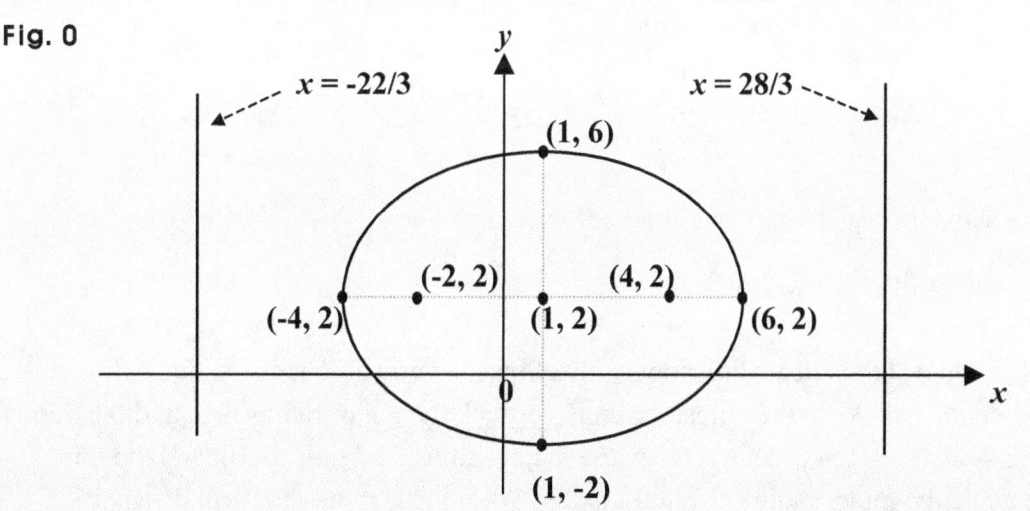

The ellipse is: $\dfrac{(x-1)^2}{5^2} + \dfrac{(y-2)^2}{4^2} = 1$.

Suggestions or Solutions
To the Problem in the Example 3

Find all the elements of the ellipse as follows: $25(x - 1)^2 + 16(y - 2)^2 = 400$.

To begin with, we can put the ellipse give this way: $\dfrac{(x-1)^2}{4^2} + \dfrac{(y-2)^2}{5^2} = 1$.

So the ellipse is vertical, the center is $(1, 2)$, the major axis is 10, and the minor axis is 8.

Next, assuming c is the focal distance, b is the major radius, and a is the minor radius, we get: $b = 5$, $a = 4$, and $c^2 = b^2 - a^2 \Rightarrow c^2 = 25 - 16 = 9 \Rightarrow c = 3$.

So the foci are $(1, 5)$ and $(1, -1)$.   And the vertices are $(1, 7)$ and $(1, -3)$.

Next, assuming e is the eccentricity, we get: $e = c/b = 3/5$.

And next, the directrices are: $y = \pm b/e + 2 = \pm b^2/c + 2 = \pm 25/3 + 2$.

If not quite sure of the idea behind the processes above, follow the steps below:

To begin with we can put the ellipse given the way below:

$$25(x - 1)^2 + 16(y - 2)^2 = 400 \Rightarrow \frac{25}{400}(x-1)^2 + \frac{16}{400}(y-2)^2 = 1$$

$$\Rightarrow \frac{5^2}{20^2}(x-1)^2 + \frac{4^2}{20^2}(y-2)^2 \Rightarrow \frac{1}{16}(x-1)^2 + \frac{1}{25}(y-2)^2 = 1 \Rightarrow \frac{(x-1)^2}{4^2} + \frac{(y-2)^2}{5^2} = 1.$$

And the standard equation of an ellipse centered at (u, v) is: $\dfrac{(x-u)^2}{a^2} + \dfrac{(y-v)^2}{b^2} = 1$.

So the center of the ellipse given is $(1, 2)$.

Next, if $b > a > 0$ in the equation above, the ellipse is vertical, b is the major radius, and a is the minor radius. So the ellipse given is <u>vertical</u>, the major radius is 5, and the minor radius is 4. And the major axis is twice the major radius, and thus, is 10, and the minor axis is twice the minor radius, and thus, is 8.   What then, about the focal distance?

Assuming c is the focal distance, b is the major radius, and a is the minor radius, we get: $c^2 = b^2 - a^2$.   So we get: $c^2 = b^2 - a^2 = 5^2 - 4^2 = 25 - 16 = 9 \Rightarrow c = 3$.

Next, the center, foci, and vertices are all in the major axis.
So if the ellipse is <u>vertical</u>, the center, foci, and vertices share the <u>same *x*-coordinate</u>.

The center is the midpoint between the foci, and the focal distance is the distance from the center to each focus, and is *c*, which is 3.

And also, the center is the midpoint between the <u>vertices</u>, too, which are the <u>endpoints</u> of the <u>major axis</u>, which is twice the major radius, which is the distance from the center to each vertex, and in this case, is *b*, which is 5.

So since the center is (1, 2), and the ellipse is vertical, the two foci are (0 + 1, 3 + 2) and (0 + 1, -3 + 2), that is, **(1, 5)** and **(1, -1)**, and next, the two vertices are (0 + 1, 5 + 2) and (0 + 1, -5 + 2), that is, **(1, 7)** and **(1, -3)**.

Next, the eccentricity of an ellipse is a ratio, <u>the focal distance over the major radius</u>.
So assuming *e* is the eccentricity, we get: ***e* = *c*/*b* = 3/5**.

And next, an ellipse has two lines called the directrices, and the distance from each to the center is a ratio, <u>the major radius over the eccentricity</u>.   So since the center is (1, 2), and the ellipse is <u>vertical</u>, the directrices are: ***y* = ±*b*/*e* + 2 = ±*b²*/*c* + 2 = ±25/3 + 2**.

Fig. 0

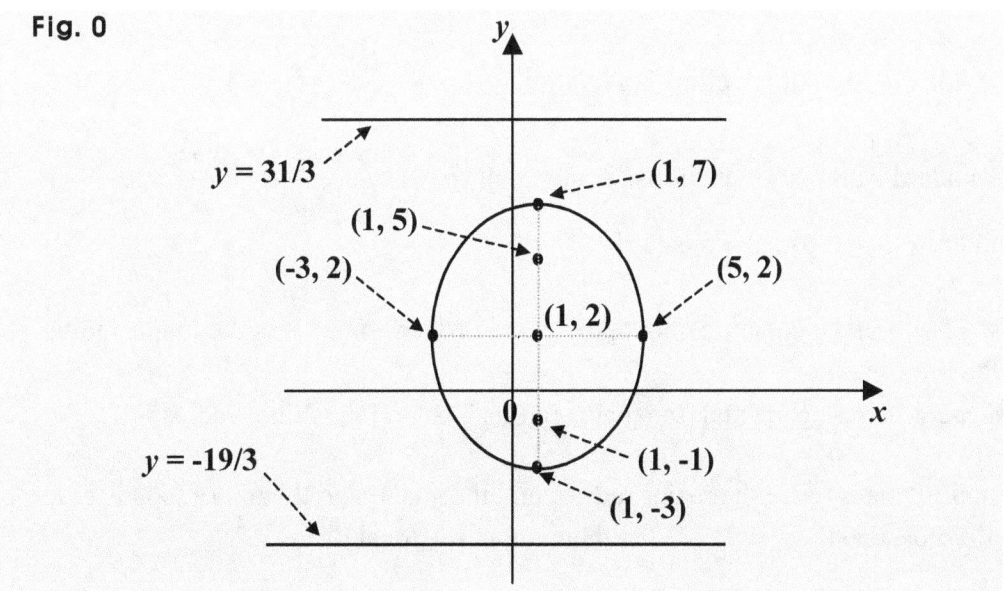

The ellipse is: $\dfrac{(x-1)^2}{4^2} + \dfrac{(y-2)^2}{5^2} = 1$.

Suggestions or Solutions
To the **Problem** in the Example **4**

Find all the elements of the ellipse as follows: $\dfrac{(x-1)^2}{4}+y^2=1$.

To begin with, the ellipse is horizontal, the center is (1, 0), the major axis is 4, and the minor axis is 2.

Next, assuming c is the focal distance, a is the major radius, and b is the minor radius, we get: $a = 2$, $b = 1$, and $c^2 = a^2 - b^2 \Rightarrow c^2 = 4 - 1 = 3 \Rightarrow c = \sqrt{3}$.

So the focal distance is $\sqrt{3}$, and the foci are $(1-\sqrt{3},0)$ and $(1+\sqrt{3},0)$.
And the vertices are (-1, 0) and (3, 0).
Next, assuming e is the eccentricity, we get: $e = c/a = \frac{\sqrt{3}}{2}$.
And next, the directrices are: $x = \pm a/e + 1 = \pm a^2/c + 1 = \pm \frac{4\sqrt{3}}{3}+1$.

If not quite sure of the idea behind the processes above, follow the steps below:

To begin with, we can put the ellipse given this way: $\dfrac{(x-1)^2}{2^2}+\dfrac{y^2}{1^2}=1$.

And the standard equation of an ellipse centered at (*u*, 0) is: $\dfrac{(x-u)^2}{a^2}+\dfrac{y^2}{b^2}=1$.
So the center of the ellipse given is (1, 0).

Next, if $a > b > 0$ in the equation above, the ellipse is horizontal, a is the major radius, and b is the minor radius.
So the ellipse given is <u>horizontal</u>, the major radius is 2, and the minor radius is 1.

And the major axis is twice the major radius, and thus, is 4, and the minor axis is twice the minor radius, and thus, is 2.   What then, about the focal distance?

Assuming c is the focal distance, a is the major radius, and b is the minor radius, we get: $c^2 = a^2 - b^2$.   So we get: $c^2 = a^2 - b^2 = 2^2 - 1^2 = 4 - 1 = 3 \Rightarrow c = \sqrt{3}$.

Next, the center, foci, and vertices are all in the major axis.
So if the ellipse is <u>horizontal</u>, the center, foci, and vertices share the <u>same y-coordinate</u>.

The center is the midpoint between the foci, and the focal distance is the distance from the center to each focus, and is *c*, which is $\sqrt{3}$.

And also, the center is the midpoint between the <u>vertices</u>, too, which are the <u>endpoints</u> of the <u>major axis</u>, which is twice the major radius, which is the distance from the center to each vertex, and in this case, is *a*, which is 2.

So since the center is (1, 0), and the ellipse is horizontal, the two foci are $(\sqrt{3}+1,0)$ and $(-\sqrt{3}+1,0),$ and the vertices are (2 + 1, 0) and (-2 + 1, 0), that is, **(3, 0)** and **(-1, 0)**.

Next, the eccentricity of an ellipse is a ratio, <u>the focal distance over the major radius</u>. So assuming *e* is the eccentricity, we get: $e = c/a = \frac{\sqrt{3}}{2}.$

And next, an ellipse has two lines called the directrices, and the distance from each to the center is a ratio, <u>the major radius over the eccentricity</u>. So since the center is (1, 0), and it is <u>horizontal</u>, the directrices are: $x = \pm a/e + 1 = \pm a^2/c + 1 = \pm \frac{4}{\sqrt{3}} + 1 = \pm \frac{4\sqrt{3}}{3} + 1.$

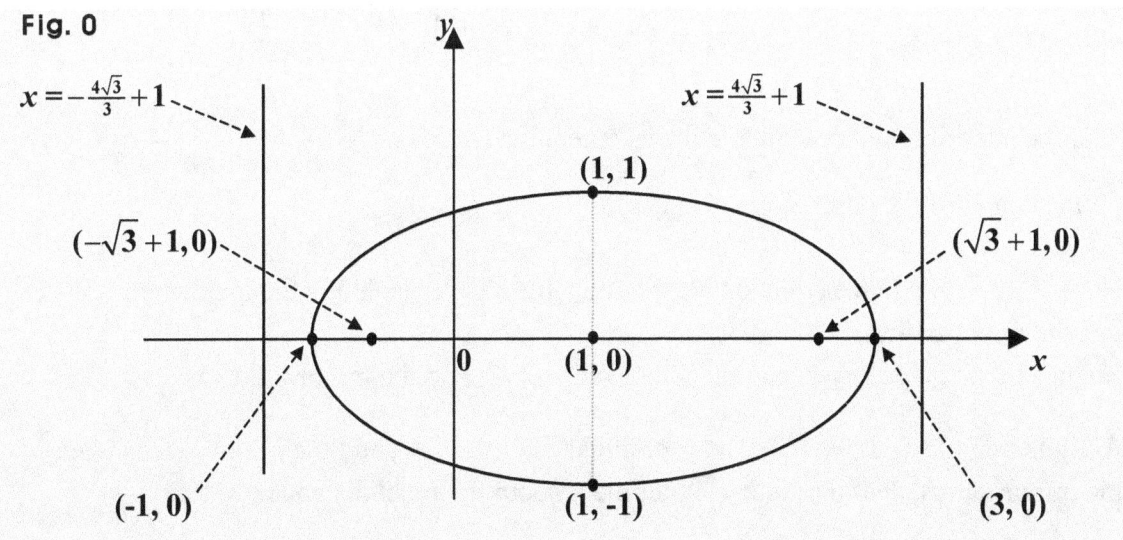

Fig. 0

The ellipse is: $\dfrac{(x-1)^2}{4} + y^2 = 1.$

Suggestions or Solutions
To the **Problem** in the Example **5**

Find all the elements of the ellipse as follows: $4(x-1)^2 + (y-2)^2 = 4$.

To begin with, we can put the ellipse give this way: $\dfrac{(x-1)^2}{1^2} + \dfrac{(y-2)^2}{2^2} = 1$.

So the ellipse is vertical, the center is $(1, 2)$, the major axis is 4, and the minor axis is 2.

Next, assuming c is the focal distance, a is the major radius, and b is the minor radius, we get: $b = 2$, $a = 1$, and $c^2 = b^2 - a^2 \Rightarrow c^2 = 4 - 1 = 3 \Rightarrow c = \sqrt{3}$.

So the foci are $(1, 2 - \sqrt{3})$ and $(1, 2 + \sqrt{3})$. And the vertices are $(\mathbf{1, 4})$ and $(\mathbf{1, 0})$.

Next, assuming e is the eccentricity, we get: $e = c/b = \frac{\sqrt{3}}{2}$.

And next, the directrices are: $y = \pm b/e + 2 = \pm b^2/c + 2 = \pm \frac{4\sqrt{3}}{3} + 2$.

If not quite sure of the idea behind the processes above, follow the steps below:

To begin with we can put the ellipse given the way below:

$$4(x-1)^2 + (y-2)^2 = 4 \Rightarrow (x-1)^2 + \frac{1}{4}(y-2)^2 = 1 \Rightarrow \frac{(x-1)^2}{1^2} + \frac{(y-2)^2}{2^2} = 1.$$

And the standard equation of an ellipse centered at $(\boldsymbol{u}, \boldsymbol{v})$ is: $\dfrac{(x-u)^2}{a^2} + \dfrac{(y-v)^2}{b^2} = 1$.

So the center of the ellipse given is $(1, 2)$.

Next, if $\boldsymbol{b > a > 0}$ in the equation above, the ellipse is vertical, $\boldsymbol{b}$ is the major radius, and $\boldsymbol{a}$ is the minor radius.

So the ellipse given is vertical, the major radius is 2, and the minor radius is 1.

And the major axis is twice the major radius, and thus, is 4, and the minor axis is twice the minor radius, and thus, is 2. What then, about the focal distance?

Assuming c is the focal distance, b is the major radius, and a is the minor radius, we get: $c^2 = b^2 - a^2$. So we get: $c^2 = b^2 - a^2 = 2^2 - 1^2 = 4 - 1 = 3 \Rightarrow c = \sqrt{3}$.

Next, the center, foci, and vertices are all in the major axis.
So if the ellipse is <u>vertical</u>, the center, foci, and vertices share the <u>same *x*-coordinate</u>.

The center is the midpoint between the foci, and the focal distance is the distance from the center to each focus, and is *c*, which is $\sqrt{3}$.

And also, the center is the midpoint between the <u>vertices</u>, too, which are the <u>endpoints</u> of the <u>major axis</u>, which is twice the major radius, which is the distance from the center to each vertex, and in this case, is *b*, which is 2.

So since the center is (1, 2), and the ellipse is vertical, the two foci are $(0+1, 2+\sqrt{3})$ and $(0+1, 2-\sqrt{3})$, that is, $(1, 2+\sqrt{3})$ and $(1, 2-\sqrt{3})$, and the two vertices are (0 + 1, 2 + 2) and (0 + 1, -2 + 2), that is, **(1, 4)** and **(1, 0)**.

Next, the eccentricity of an ellipse is a ratio, <u>the focal distance over the major radius</u>. So assuming *e* is the eccentricity, we get: $e = c/b = \frac{\sqrt{3}}{2}$.

And next, an ellipse has two lines called the directrices, and the distance from each to the center is a ratio, <u>the major radius over the eccentricity</u>. So since the center is (1, 2), and it is <u>vertical</u>, the directrices are: $y = \pm b/e + 2 = \pm b^2/c + 2 = \pm\frac{4}{\sqrt{3}} + 2 = \pm\frac{4\sqrt{3}}{3} + 2$.

Fig. 0

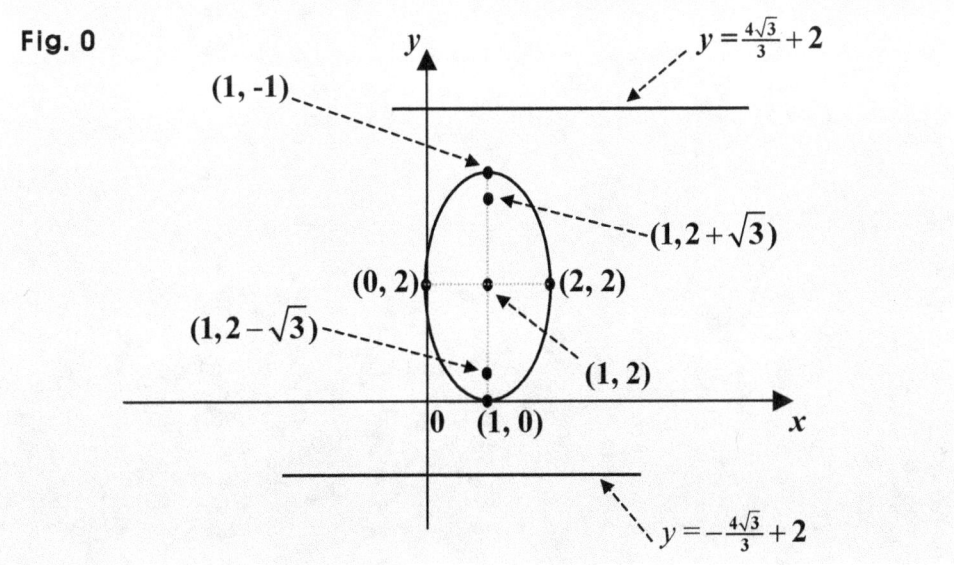

The ellipse is: $\dfrac{(x-1)^2}{1^2} + \dfrac{(y-2)^2}{2^2} = 1$.

Examples 4 in Ellipses

Find if each ellipse below is horizontal or vertical, and the center, foci, vertices, major axis, minor axis, eccentricity, and directrices.

0. $25(x-2)^2 + 16(y-1)^2 = 1$

1. $4x^2 + 8x + 4 + y^2 = 1$

2. $x^2 + 8y + 4 + 4y^2 = 1$

3. $16x^2 + 32x + 25y^2 + 100y = 284$

4. $12x - 3x^2 - 8y - 4y^2 - 4 = 0$

Suggestions or Solutions
To the Problem in the Example 0

Find all the elements of the ellipse as follows: $25(x - 1)^2 + 16(y - 2)^2 = 1$.

To begin with, we can put the ellipse give this way: $\dfrac{(x-1)^2}{(\frac{1}{5})^2} + \dfrac{(y-2)^2}{(\frac{1}{4})^2} = 1$.

So the ellipse is vertical, the center is $(1, 2)$, the major axis is $\frac{1}{2}$, and the minor axis is $\frac{2}{5}$.

Next, assuming c is the focal distance, b is the major radius, and a is the minor radius, we get: $b = \frac{1}{4}, a = \frac{1}{5}$, and $c^2 = b^2 - a^2 \Rightarrow c^2 = (\frac{1}{4})^2 - (\frac{1}{5})^2 = \frac{9}{400} \Rightarrow c = \frac{3}{20}$.

So the foci are $(1, \frac{43}{20})$ and $(1, \frac{37}{20})$. And the vertices are $(1, \frac{9}{4})$ and $(2, \frac{7}{4})$.

Next, assuming e is the eccentricity, we get: $e = c/b = \frac{3}{5}$.

And next, the directrices are: $y = \pm b/e + 2 = \pm b^2/c + 2 = \pm \frac{5}{12} + 2$.

If not quite sure of the idea behind the processes above, follow the steps below:

To begin with, we can put the ellipse given this way: $\dfrac{(x-1)^2}{(\frac{1}{5})^2} + \dfrac{(y-2)^2}{(\frac{1}{4})^2} = 1$.

And the standard equation of an ellipse centered at (u, v) is: $\dfrac{(x-u)^2}{a^2} + \dfrac{(y-v)^2}{b^2} = 1$.

So the center of the ellipse given is $(1, 2)$.

Next, if $b > a > 0$ in the equation above, the ellipse is vertical, b is the major radius, and a is the minor radius.
So the ellipse given is <u>vertical</u>, the major radius is $\frac{1}{4}$, and the minor radius is $\frac{1}{5}$.

And the major axis is twice the major radius, and thus, is $\frac{1}{2}$, and the minor axis is twice the minor radius, and thus, is $\frac{2}{5}$. What then, about the focal distance?

Assuming c is the focal distance, b is the major radius, and a is the minor radius, we get: $b = \frac{1}{4}, a = \frac{1}{5}$, and $c^2 = b^2 - a^2 \Rightarrow c^2 = (\frac{1}{4})^2 - (\frac{1}{5})^2 = \frac{1}{16} - \frac{1}{25} = \frac{9}{400} \Rightarrow c = \frac{3}{20}$.

Next, the center, foci, and vertices are all in the major axis.
So if the ellipse is <u>vertical</u>, the center, foci, and vertices share the <u>same *x*-coordinate</u>.

The center is the midpoint between the foci, and the focal distance is the distance from the center to each focus, and is *c*, which is $\frac{3}{20}$.

And also, the center is the midpoint between the <u>vertices</u>, too, which are the <u>endpoints</u> of the <u>major axis</u>, which is twice the major radius, which is the distance from the center to each vertex, and in this case, is *b*, which is 5.

So since the center is (1, 2), and the ellipse is vertical, the foci are $(0+1, \frac{3}{20}+2)$ and $(0+1, -\frac{3}{20}+2)$, that is, $(1, \frac{43}{20})$ and $(1, \frac{37}{20})$, and the vertices are $(0+1, \frac{1}{4}+2)$ and $(0+1, -\frac{1}{4}+2)$, that is, $(1, \frac{9}{4})$ and $(1, \frac{7}{4})$.

Next, the eccentricity of an ellipse is a ratio, <u>the focal distance over the major radius</u>. So assuming *e* is the eccentricity, we get: $e = c/b = \frac{3}{20} \cdot 4 = \frac{3}{5}$.

And next, an ellipse has two lines called the directrices, and the distance from each to the center is a ratio, <u>the major radius over the eccentricity</u>.   So since the center is (1, 2), and it is <u>vertical</u>, the directrices are: $y = \pm b/e + 2 = \pm b^2/c + 2 = \pm \frac{1}{16} \cdot \frac{20}{3} + 2 = \pm \frac{5}{12} + 2$.

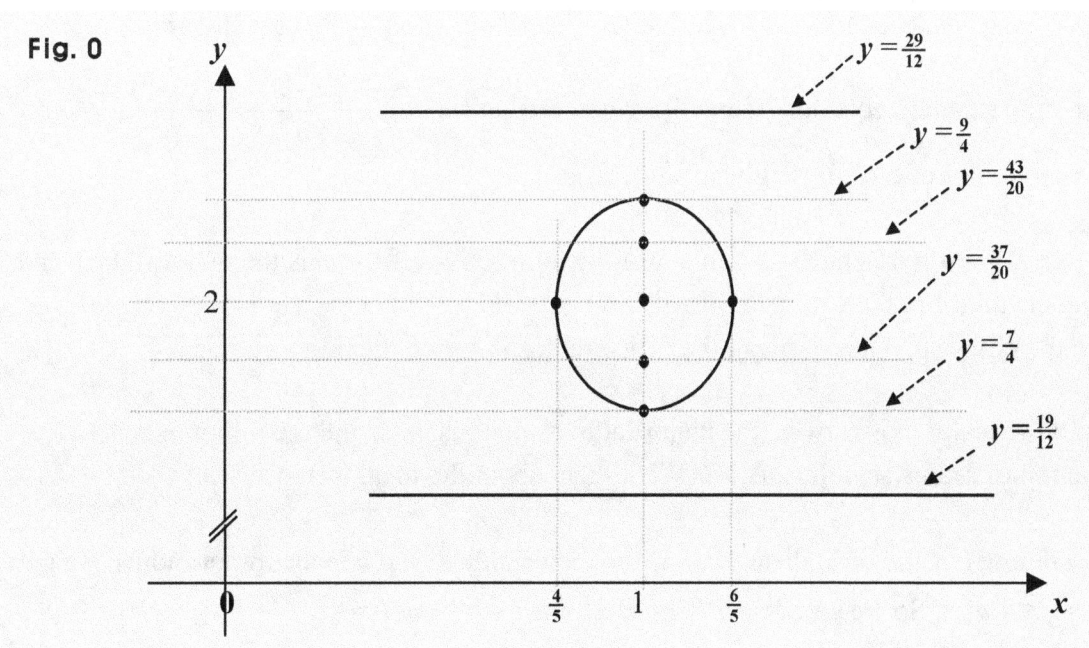

Fig. 0

Suggestions or Solutions
To the Problem in the Example 1

Find all the elements of the ellipse as follows: $4x^2 + 8x + 4 + y^2 = 1$.

To begin with, we can put the ellipse give this way: $\dfrac{(x+1)^2}{(\frac{1}{2})^2} + \dfrac{y^2}{1^2} = 1$.

So the ellipse is vertical, the center is (-1, 0), the major axis is 2, and the minor axis is 1.

Next, assuming c is the focal distance, a is the major radius, and b is the minor radius, we get: $b = 1$, $a = \frac{1}{2}$, and $c^2 = b^2 - a^2 \Rightarrow c^2 = 1 - \frac{1}{4} = \frac{3}{4} \Rightarrow c = \frac{\sqrt{3}}{2}$.

So the foci are $\left(-1, \frac{\sqrt{3}}{2}\right)$ and $\left(-1, -\frac{\sqrt{3}}{2}\right)$. And the vertices are (-1, 1) and (-1, -1).

Next, assuming e is the eccentricity, we get: $e = c/b = \frac{\sqrt{3}}{2}$.

And next, the directrices are: $y = \pm b/e = \pm b^2/c = \pm \frac{2\sqrt{3}}{3}$.

If not quite sure of the idea behind the processes above, follow the steps below:

To begin with we can put the ellipse given the way below:

$$4x^2 + 8x + 4 + y^2 = 4(x^2 + 2x + 1) + y^2 = 4(x + 1)^2 + y^2 = 1 \Rightarrow \dfrac{(x+1)^2}{(\frac{1}{2})^2} + \dfrac{y^2}{1^2} = 1.$$

And the standard equation of an ellipse centered at (u, v) is: $\dfrac{(x-u)^2}{a^2} + \dfrac{(y-v)^2}{b^2} = 1$.

So the center of the ellipse given is (-1, 0).

Next, if $b > a > 0$ in the equation above, the ellipse is vertical, b is the major radius, and a is the minor radius.

So the ellipse given is <u>vertical</u>, the major radius is 1, and the minor radius is $\frac{1}{2}$.

And the major axis is twice the major radius, and thus, is 2, and the minor axis is twice the minor radius, and thus, is 1. What then, about the focal distance?

Assuming c is the focal distance, b is the major radius, and a is the minor radius, we get: $c^2 = b^2 - a^2$. So we get: $c^2 = b^2 - a^2 \Rightarrow c^2 = 1 - \frac{1}{4} = \frac{3}{4} \Rightarrow c = \frac{\sqrt{3}}{2}$.

Next, the center, foci, and vertices are all in the major axis.
So if the ellipse is <u>vertical</u>, the center, foci, and vertices share the <u>same *x*-coordinate</u>.

The center is the midpoint between the foci, and the focal distance is the distance from the center to each focus, and is *c*, which is $\frac{\sqrt{3}}{2}$.

And also, the center is the midpoint between the <u>vertices</u>, too, which are the <u>endpoints</u> of the <u>major axis</u>, which is twice the major radius, which is the distance from the center to each vertex, and in this case, is *b*, which is 1.

So since the center is (-1, 0), and the ellipse is vertical, the two foci are $\mathbf{(0-1,\frac{\sqrt{3}}{2})}$ and $\mathbf{(0-1,-\frac{\sqrt{3}}{2})}$, that is, $\mathbf{(-1,\frac{\sqrt{3}}{2})}$ and $\mathbf{(-1,-\frac{\sqrt{3}}{2})}$, and the vertices are (0 – 1, 1) and (0 – 1, -1), that is, **(-1, 1)** and **(-1, -1)**.

Next, the eccentricity of an ellipse is a ratio, <u>the focal distance over the major radius</u>. So assuming *e* is the eccentricity, we get: $\boldsymbol{e = c/b = \frac{\sqrt{3}}{2}}$.

And next, an ellipse has two lines called the directrices, and the distance from each to the center is a ratio, <u>the major radius over the eccentricity</u>.   So since the center is (-1, 0), and the ellipse is <u>vertical</u>, the directrices are: $\boldsymbol{y = \pm b/e = \pm b^2/c = \pm\frac{2}{\sqrt{3}} = \pm\frac{2\sqrt{3}}{3}}$.

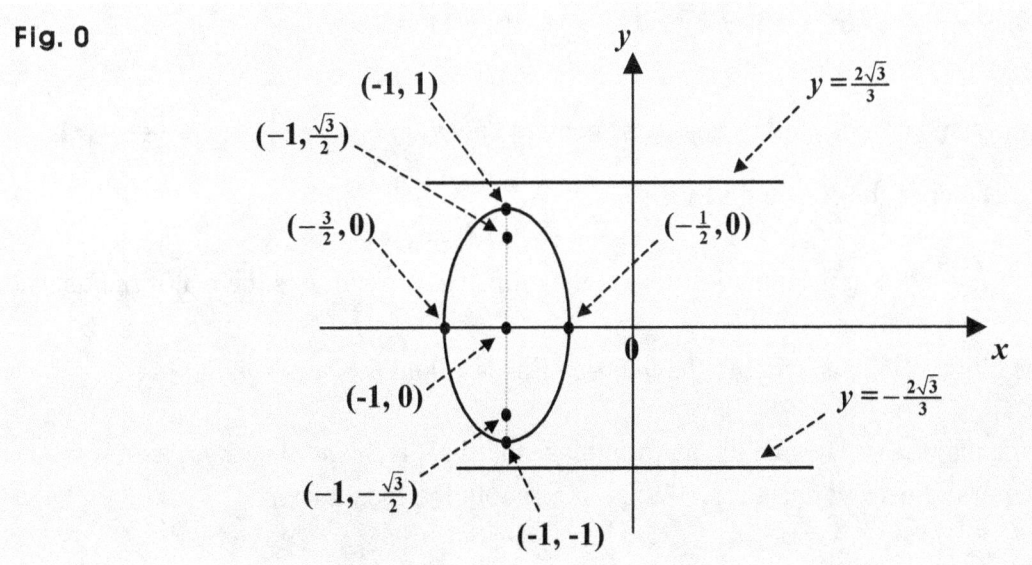

Fig. 0

The ellipse is: $\dfrac{(x+1)^2}{(\frac{1}{2})^2} + \dfrac{y^2}{1^2} = 1.$

Suggestions or Solutions
To the Problem in the Example 2

Find all the elements of the ellipse as follows: $x^2 + 8y + 4 + 4y^2 = 1$.

To begin with, we can put the ellipse give this way: $\dfrac{x^2}{1^2} + \dfrac{(y+1)^2}{(\frac{1}{2})^2} = 1$.

So the ellipse is horizontal, the center is (0, -1), the major axis is 2, and the minor axis is 1. Next, assuming c is the focal distance, a is the major radius, and b is the minor radius, we get: $a = 1$, $b = \frac{1}{2}$, and $c^2 = a^2 - b^2 \Rightarrow c^2 = 1 - \frac{1}{4} = \frac{3}{4} \Rightarrow c = \frac{\sqrt{3}}{2}$.

So the foci are $(\frac{\sqrt{3}}{2}, -1)$ and $(-\frac{\sqrt{3}}{2}, -1)$. And the vertices are (-1, -1) and (1, -1).

Next, assuming e is the eccentricity, we get: $e = c/a = \frac{\sqrt{3}}{2}$.

And next, the directrices are: $x = \pm a/e = \pm a^2/c = \pm \frac{2\sqrt{3}}{3}$.

If not quite sure of the idea behind the processes above, follow the steps below:

To begin with we can put the ellipse given the way below:

$x^2 + 8y + 4 + 4y^2 = x^2 + 4(y^2 + 2y + 1) = x^2 + 4(y + 1)^2 = 1 \Rightarrow \dfrac{x^2}{1^2} + \dfrac{(y+1)^2}{(\frac{1}{2})^2} = 1$.

And the standard equation of an ellipse centered at **(u, v)** is: $\dfrac{(x-u)^2}{a^2} + \dfrac{(y-v)^2}{b^2} = 1$.

So the center of the ellipse given is (0, -1).

Next, if **b > a > 0** in the equation above, the ellipse is vertical, **b** is the major radius, and **a** is the minor radius.

So the ellipse given is <u>vertical</u>, the major radius is 1, and the minor radius is $\frac{1}{2}$.

And the major axis is twice the major radius, and thus, is 2, and the minor axis is twice the minor radius, and thus, is 1.   What then, about the focal distance?

Assuming **c** is the focal distance, **a** is the major radius, and **b** is the minor radius, we get: $c^2 = a^2 - b^2$.   So we get: $c^2 = a^2 - b^2 \Rightarrow c^2 = 1 - \frac{1}{4} = \frac{3}{4} \Rightarrow c = \frac{\sqrt{3}}{2}$.

Next, the center, foci, and vertices are all in the major axis.
So if the ellipse is <u>horizontal</u>, the center, foci, and vertices share the <u>same y-coordinate</u>.

The center is the midpoint between the foci, and the focal distance is the distance from the center to each focus, and is c, which is $\frac{\sqrt{3}}{2}$.

And also, the center is the midpoint between the <u>vertices</u>, too, which are the <u>endpoints</u> of the <u>major axis</u>, which is twice the major radius, which is the distance from the center to each vertex, and in this case, is a, which is 1.

So since the center is (0, -1), and the ellipse is horizontal, the two foci are $(\frac{\sqrt{3}}{2}, 0-1)$ and $(-\frac{\sqrt{3}}{2}, 0-1)$, that is, $(\frac{\sqrt{3}}{2}, -1)$ and $(-\frac{\sqrt{3}}{2}, -1)$, and the vertices are (1, 0 – 1) and (-1, 0 – 1), that is, **(1, -1) and (-1, -1)**.

Next, the eccentricity of an ellipse is a ratio, <u>the focal distance over the major radius</u>. So assuming e is the eccentricity, we get: $e = c/a = \frac{\sqrt{3}}{2}$.

And next, an ellipse has two lines called the directrices, and the distance from each to the center is a ratio, <u>the major radius over the eccentricity</u>.   So since the center is (0, -1), and the ellipse is <u>horizontal</u>, the directrices are: $x = \pm a/e = \pm a^2/c = \pm \frac{2}{\sqrt{3}} = \pm \frac{2\sqrt{3}}{3}$.

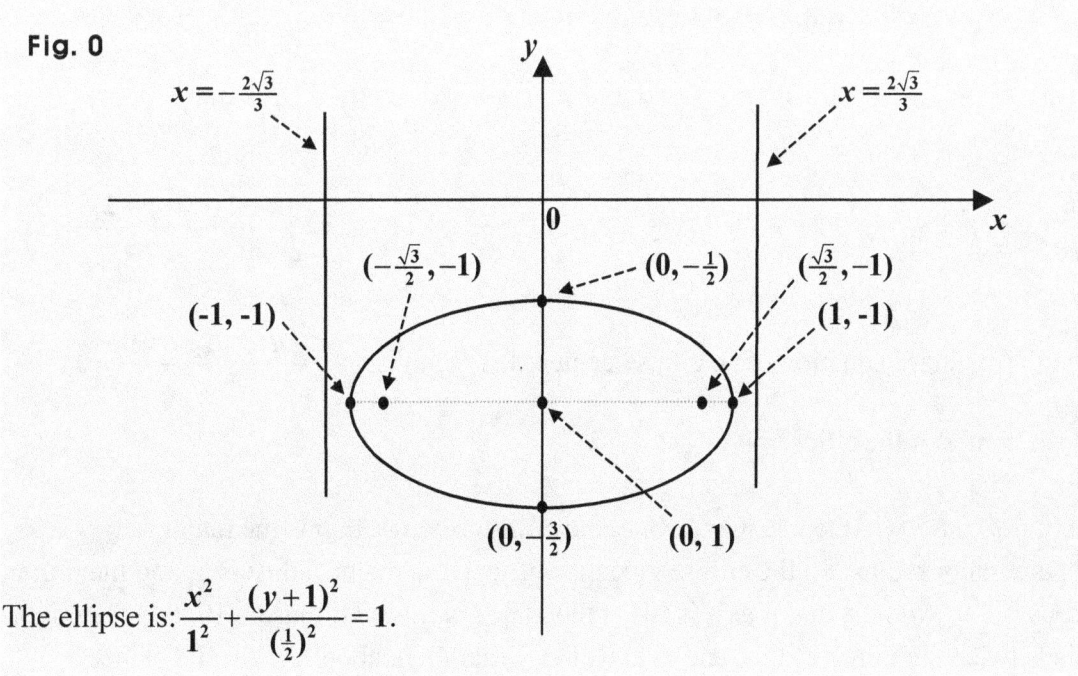

Fig. 0

$x = -\frac{2\sqrt{3}}{3}$ $x = \frac{2\sqrt{3}}{3}$

$(-\frac{\sqrt{3}}{2}, -1)$ $(0, -\frac{1}{2})$ $(\frac{\sqrt{3}}{2}, -1)$

(-1, -1)   (1, -1)

$(0, -\frac{3}{2})$ (0, 1)

The ellipse is: $\dfrac{x^2}{1^2} + \dfrac{(y+1)^2}{(\frac{1}{2})^2} = 1.$

Suggestions or Solutions
To the Problem in the Example 3

Find all the elements of the ellipse as follows: $16x^2 + 32x + 25y^2 + 100y = 284$.

To begin with, we can put the ellipse give this way: $\dfrac{(x+1)^2}{5^2} + \dfrac{(y+2)^2}{4^2} = 1$.

So the ellipse is horizontal, the center is (-1, -2), the major axis is 2, and the minor axis is 1. Next, assuming c is the focal distance, a is the major radius, and b is the minor radius, we get: $a = 5$, $b = 4$, and $c^2 = a^2 - b^2 \Rightarrow c^2 = 25 - 16 = 3 \Rightarrow c = 3$.

So the foci are (2, -2) and (-4, -2).   And the vertices are (4, -2) and (-6, -2).

Next, assuming e is the eccentricity, we get: $e = c/a = 3/5$.

And next, the directrices are: $x = \pm a/e - 1 = \pm a^2/c - 1 = \pm 25/3 - 1$.

If not quite sure of the idea behind the processes above, follow the steps below:

To begin with we can put the ellipse given the way below:

$16x^2 + 32x + 25y^2 + 100y - 284 = 16(x^2 + 2x + 1 - 1) + 25(y^2 + 4y + 4 - 4) - 284$

$= 16(x + 1)^2 - 16 + 25(y + 2)^2 - 100 - 284 = 16(x + 1)^2 + 25(y + 2)^2 - 400 = 0 \Rightarrow$

$\dfrac{16}{400}(x+1)^2 + \dfrac{25}{400}(y+2)^2 - 1 = 0 \Rightarrow \dfrac{1}{25}(x+1)^2 + \dfrac{1}{16}(y+2)^2 = 1 \Rightarrow \dfrac{(x+1)^2}{5^2} + \dfrac{(y+2)^2}{4^2} = 1$.

And the standard equation of an ellipse centered at (u, v) is: $\dfrac{(x-u)^2}{a^2} + \dfrac{(y-v)^2}{b^2} = 1$.

So the center of the ellipse given is (-1, -2).

Next, if $b > a > 0$ in the equation above, the ellipse is vertical, b is the major radius, and a is the minor radius. So the ellipse given is vertical, the major radius is 5, and the minor radius is 4.   And the major axis is twice the major radius, and thus, is 10, and the minor axis is twice the minor radius, and thus, is 8.   What then, about the focal distance?

Assuming c is the focal distance, a is the major radius, and b is the minor radius, we get: $c^2 = a^2 - b^2$.   So we get: $c^2 = a^2 - b^2 \Rightarrow c^2 = 25 - 16 = 9 \Rightarrow c = 3$.

Next, the center, foci, and vertices are all in the major axis.
So if the ellipse is <u>horizontal</u>, the center, foci, and vertices share the <u>same y-coordinate</u>.

The center is the midpoint between the foci, and the focal distance is the distance from the center to each focus, and is c, which is 3.

And also, the center is the midpoint between the <u>vertices</u>, too, which are the <u>endpoints</u> of the <u>major axis</u>, which is twice the major radius, which is the distance from the center to each vertex, and in this case, is a, which is 5.

So since the center is (-1, -2), and the ellipse is horizontal, the two foci are $(-3 - 1, 0 - 2)$ and $(3 - 1, 0 - 2)$, that is, **(-4, -2)** and **(2, -2)**, and next, the two vertices are $(5 - 1, 0 - 2)$ and $(-5 - 1, 0 - 2)$, that is, **(4, -2)** and **(-6, -2)**.

Next, the eccentricity of an ellipse is a ratio, <u>the focal distance over the major radius</u>. So assuming e is the eccentricity, we get: $e = c/a = 3/5$.

And next, an ellipse has two lines called the directrices, and the distance from each to the center is a ratio, <u>the major radius over the eccentricity</u>. So since the center is (-1, -2), and the ellipse is <u>horizontal</u>, the directrices are: $x = \pm a/e - 1 = \pm a^2/c - 1 = \pm 25/3 - 1$.

Fig. 0

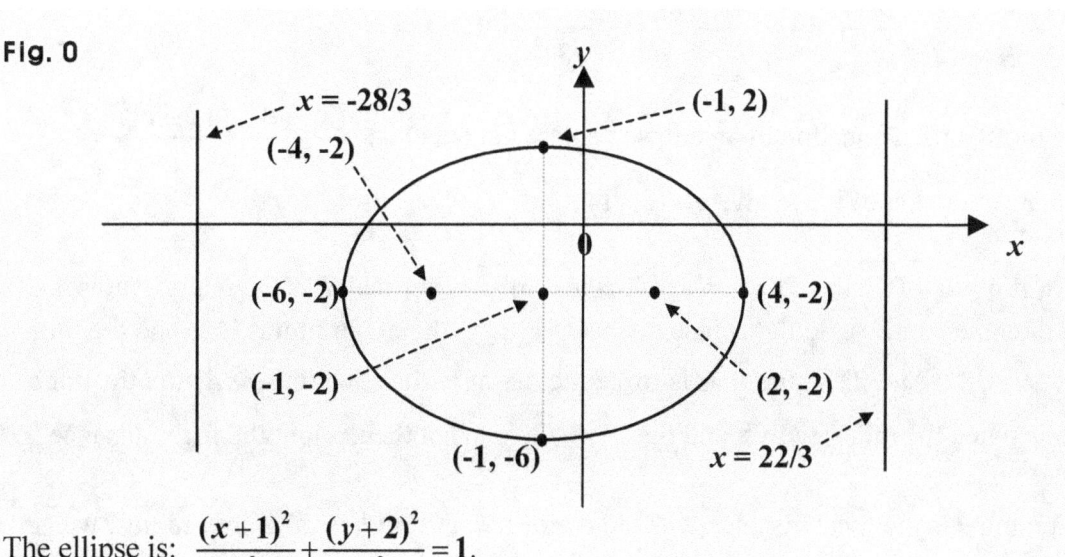

The ellipse is: $\dfrac{(x+1)^2}{5^2} + \dfrac{(y+2)^2}{4^2} = 1$.

Suggestions or Solutions
To the Problem in the Example 4

Find all the elements of the ellipse as follows: $12x - 3x^2 - 8y - 4y^2 - 4 = 0$.

To begin with, we can put the ellipse give this way: $\dfrac{(x-2)^2}{2^2} + \dfrac{(y+1)^2}{(\sqrt{3})^2} = 1$.

So the ellipse is horizontal, the center is (2, -1), the major axis is 4, and the minor axis is $2\sqrt{3}$. Next, assuming c is the focal distance, a is the major radius, and b is the minor radius, we get: $a = 2$, $b = \sqrt{3}$, and $c^2 = a^2 - b^2 \Rightarrow c^2 = 4 - 3 = 1 \Rightarrow c = 1$.

So the foci are (1, -1) and (3, -1). And the vertices are (4, -1) and (0, -1).
Next, assuming e is the eccentricity, we get: $e = c/a = \frac{1}{2}$.
And next, the directrices are: $x = \pm a/e + 2 = \pm a^2/c + 2 = \pm 4 + 2$.

If not quite sure of the idea behind the processes above, follow the steps below:

To begin with we can put the ellipse given the way below:
$12x - 3x^2 - 8y - 4y^2 - 4 = 0 \Rightarrow 3x^2 - 12x + 4y^2 + 8y + 4 = 0$
$\Rightarrow 3(x^2 - 4x + 4 - 4) + 4(y^2 + 2y + 1) = 3(x-2)^2 + 4(y+1)^2 - 12 = 0$
$\Rightarrow \dfrac{(x-2)^2}{4} + \dfrac{(y+1)^2}{3} = 1 \Rightarrow \dfrac{(x-2)^2}{2^2} + \dfrac{(y+1)^2}{(\sqrt{3})^2} = 1$.

And the standard equation of an ellipse centered at (u, v) is: $\dfrac{(x-u)^2}{a^2} + \dfrac{(y-v)^2}{b^2} = 1$.
So the center of the ellipse given is (2, -1).

Next, if $b > a > 0$ in the equation above, the ellipse is vertical, b is the major radius, and a is the minor radius. So the ellipse given is <u>vertical</u>, the major radius is 2, and the minor radius is $\sqrt{3}$. And the major axis is twice the major radius, and thus, is 4, and the minor axis is twice the minor radius, and thus, is $2\sqrt{3}$. What then, about the focal distance?

Assuming c is the focal distance, a is the major radius, and b is the minor radius, we get: $c^2 = a^2 - b^2$. So we get: $c^2 = a^2 - b^2 \Rightarrow c^2 = 4 - 3 = 1 \Rightarrow c = 1$.

Next, the center, foci, and vertices are all in the major axis.
So if the ellipse is <u>horizontal</u>, the center, foci, and vertices share the <u>same *y*-coordinate</u>.

The center is the midpoint between the foci, and the focal distance is the distance from the center to each focus, and is *c*, which is 1.

And also, the center is the midpoint between the <u>vertices</u>, too, which are the <u>endpoints</u> of the <u>major axis</u>, which is twice the major radius, which is the distance from the center to each vertex, and in this case, is *a*, which is 2.

So since the center is (2, -1), and the ellipse is horizontal, the two foci are (-1 + 2, 0 − 1) and (1 + 2, 0 − 1), that is, **(1, -1)** and **(3, -1)**, and the two vertices are (-2 + 2, 0 − 1) and (2 + 2, 0 − 1), that is, **(0, -1)** and **(4, -1)**.

Next, the eccentricity of an ellipse is a ratio, <u>the focal distance over the major radius</u>. So assuming *e* is the eccentricity, we get: $e = c/a = \frac{1}{2}$.

And next, an ellipse has two lines called the directrices, and the distance from each to the center is a ratio, <u>the major radius over the eccentricity</u>.   So since the center is (2, -1), and the ellipse is <u>horizontal</u>, the directrices are: $x = \pm a/e + 2 = \pm a^2/c + 2 = \pm 4 + 2$.

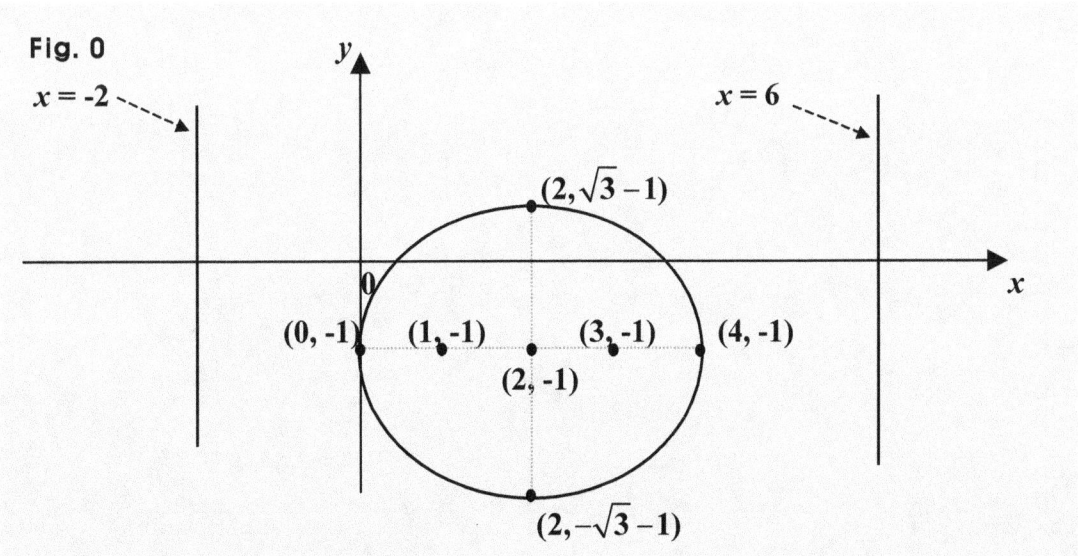

Fig. 0

$$\frac{(x-2)^2}{2^2} + \frac{(y+1)^2}{(\sqrt{3})^2} = 1.$$

Examples 5 in Ellipses

In each example below, the ellipse is in the *x-y* plane.

0.   Find the ellipse that passes through a point $(1, \frac{2\sqrt{10}}{3})$, and has foci at (0, 0) and (4, 0).

1.   Find the ellipse in which the four endpoints of the two main axes are: (1, 3), (3, 4), (5, 3), and (3, 2).

2.   Find the foci of the ellipse that passes through (-1, 1), (-2, -1), (-3, 1), and (-2, 3).

3.   Find the ellipse of which the eccentricity is 3/5, and a directrix is a line $x = 28/3$, and is corresponding to a focus at (4, 2).

4.   Suppose C is a point in a line segment $\overline{AB}$, and $\overline{AC} / \overline{CB} = 1/2$. Then, assuming the end point A is moving along the *x*-axis, and at the same time, the end point B is moving along the *y*-axis, find the curve that the point C makes.

5.   Find the maximum area of a rectangle inscribed in an ellipse $x^2 + 9y^2 = 9$.

Suggestions or Solutions
To the **Problem** in the Example **0**

Find the ellipse that passes through a point $(1, \dfrac{2\sqrt{10}}{3})$, and has foci at (0, 0) and (4, 0).

To begin with, the ellipse is horizontal, and the center is (2, 0).

So next, assuming the ellipse is: $\dfrac{(x-2)^2}{a^2} + \dfrac{y^2}{b^2} = 1$, we have: $a > b > 0$.

Next, assuming s is the sum of two distances from a point in the ellipse to the foci, we get: $s = 2a$.   And also, since the ellipse has the given point, we can get s the way below:

$$s = \sqrt{(1-0)^2 + (\tfrac{2\sqrt{10}}{3}-0)^2} + \sqrt{(1-4)^2 + (\tfrac{2\sqrt{10}}{3}-0)^2} = \sqrt{1 + \tfrac{40}{9}} + \sqrt{9 + \tfrac{40}{9}} = \tfrac{7}{3} + \tfrac{11}{3} = 6.$$

So we get: $a = 3$.

Next, assuming c is the focal distance, we get: $c = 2$, and $c^2 = a^2 - b^2 \Rightarrow b^2 = a^2 - c^2$ $\Rightarrow b^2 = 9 - 4 = 5 \Rightarrow b = \sqrt{5}$.

So the ellipse is: $\dfrac{(x-2)^2}{9} + \dfrac{y^2}{5} = 1$, often put this way, too: $\dfrac{(x-2)^2}{3^2} + \dfrac{y^2}{(\sqrt{5})^2} = 1$.

If not quite sure of the idea behind the processes above, follow the steps below:

First, the standard equation of an ellipse centered at (u, v) is: $\dfrac{(x-u)^2}{a^2} + \dfrac{(y-v)^2}{b^2} = 1$.

And if $a > b > 0$, the ellipse is horizontal, a is the major radius, and b is the minor radius. If however, $b > a > 0$, it is vertical, b is the major, and a is the minor.

And if <u>horizontal</u>, the center, foci, and vertices share the <u>same y-coordinate</u>. So since the foci given have the same y-coordinates, the ellipse we want to find is horizontal.

Next, the center is the midpoint between the foci. So the center is: {(0 + 4)/2, (0 + 0)/2}, and thus, is: (2, 0).

Next, in an ellipse, the sum of two distances from a point to the two foci is constant, and is in fact, the length of the major axis.

So assuming s is the sum of the two distances, we get: $s = 2a$.

And since the given point $(1, \dfrac{2\sqrt{10}}{3})$ is in the ellipse, the sum of the two distances from the point given to the two foci is $2a$, too.   So finding the sum s, we get:

$$s = \sqrt{(1-0)^2 + (\tfrac{2\sqrt{10}}{3}-0)^2} + \sqrt{(1-4)^2 + (\tfrac{2\sqrt{10}}{3}-0)^2} = \sqrt{1+\tfrac{40}{9}} + \sqrt{9+\tfrac{40}{9}} = \tfrac{7}{3}+\tfrac{11}{3} = 6.$$

So we get: $a = 3$.   What then, about b?

We have a connective equation $c^2 = a^2 - b^2$, where c is the focal distance, a is the major radius, and b is the minor radius.   So we get: $b^2 = a^2 - c^2$.   What then, about c?

The focal distance is the distance from the center to each focus.
So since the center is $(2, 0)$, and a focus is $(0, 0)$, we get: $c = 2$.

And thus, we get: $b^2 = a^2 - c^2 = 9 - 4 = 5 \Rightarrow b = \sqrt{5}$.

So the ellipse is: $\dfrac{(x-2)^2}{9} + \dfrac{y^2}{5} = 1$, often put this way, too: $\dfrac{(x-2)^2}{3^2} + \dfrac{y^2}{(\sqrt{5})^2} = 1$.

Fig. 0

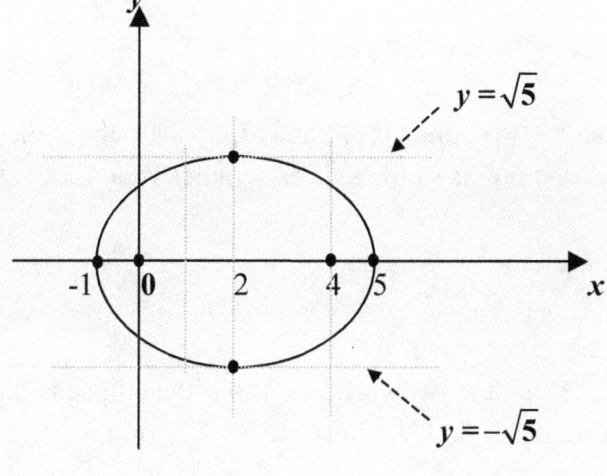

Suggestions or Solutions
To the **Problem** in the Example **1**

Find the ellipse where the four endpoints of the two main axes are: (1, 3), (4, 1), (7, 3), and (4, 5).

To begin with, taking the distance from (1, 3) to (7, 3), we get: 6.
Next, taking the distance from (4, 1) to (4, 5), we get: 4.

So the major axis is 6, the minor axis is 4, and the ellipse is horizontal.

Next, the center is (4, 3).

So the ellipse is: $\dfrac{(x-4)^2}{9} + \dfrac{(y-3)^2}{4} = 1$, often put this way, too: $\dfrac{(x-4)^2}{3^2} + \dfrac{(y-3)^2}{2^2} = 1$.

If not quite sure of the idea behind the processes above, follow the steps below:

First, the standard equation of an ellipse centered at (*u, v*) is: $\dfrac{(x-u)^2}{a^2} + \dfrac{(y-v)^2}{b^2} = 1$.

And if *a* > *b* > **0**, the ellipse is horizontal, *a* is the major radius, and *b* is the minor radius. If however, *b* > *a* > **0**, it is vertical, *b* is the major, and *a* is the minor.

Next, we can notice that the two points (4, 1) and (4, 5) share the same *x*-coordinate, so the two points are the two end points of the axis of symmetry parallel to the *y*-axis.

And taking the distance between the two, we get: 4.

Next, we can also notice that the two points (1, 3) and (7, 3) share the same *y*-coordinate, so the two points are the two end points of the axis of symmetry parallel to the *x*-axis.

And taking the distance between the two, we get: 6.

So the major axis is 6, and the minor axis is 4.
And two points (1, 3) and (7, 3) are the two vertices. And if the ellipse is horizontal, the vertices share the same *y*-coordinate. So the ellipse we want to find is horizontal.

Next, the center is the midpoint between the vertices.

So the center is: {(1 + 7)/2, (3 + 3)/2}, and thus, is (4, 3).

So the ellipse is: $\dfrac{(x-4)^2}{9}+\dfrac{(y-3)^2}{4}=1$, often put this way, too: $\dfrac{(x-4)^2}{3^2}+\dfrac{(y-3)^2}{2^2}=1$.

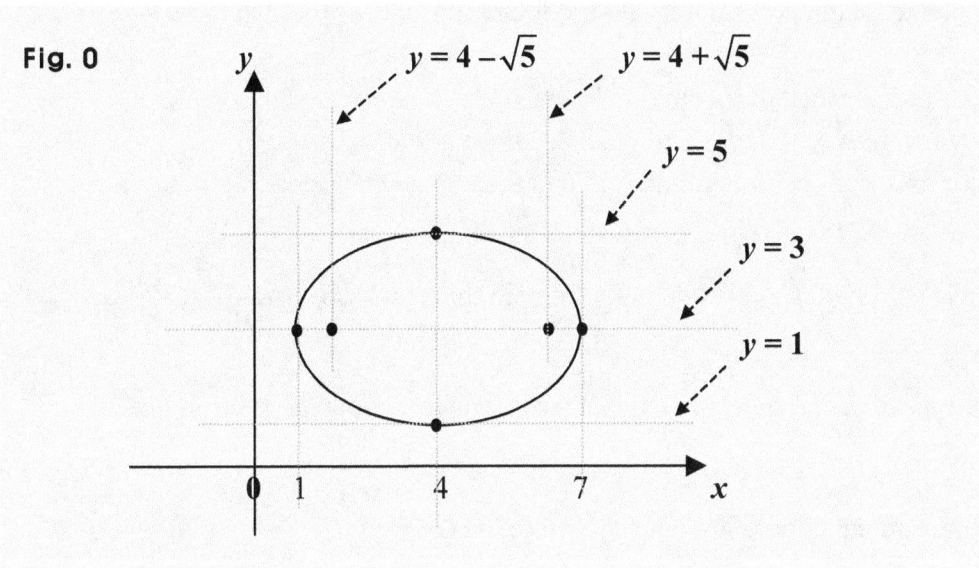

Fig. 0

Suggestions or Solutions
To the **Problem** in the Example **2**

Find the foci of the ellipse that passes through (-1, 1), (-2, -1), (-3, 1), and (-2, 3).

To begin with, taking the midpoint of (-1, 1) and (-3, 1), we get: (-2, 1).
And taking the midpoint of (-2, -1) and (-2, 3), we get: (-2, 1), too.
So we can notice that the four points are the four endpoints of the two main axes.

Thus next, taking the distance from (-1, 1) to (-3, 1), we get: 2.
Next, taking the distance from (-2, -1) to (-2, 3), we get: 4.
So the major axis is 4, the minor axis is 2, and the ellipse is vertical.
And next, the center is (-2, 1).

So the ellipse is: $(x+2)^2 + \dfrac{(y-1)^2}{4} = 1$, often put this way, too: $\dfrac{(x+2)^2}{1^2} + \dfrac{(y-1)^2}{2^2} = 1$.

If not quite sure of the idea behind the processes above, follow the steps below:

First, the standard equation of an ellipse centered at **(*u*, *v*)** is: $\dfrac{(x-u)^2}{a^2} + \dfrac{(y-v)^2}{b^2} = 1$.

And if **b > a > 0**, the ellipse is vertical, **b** is the major radius, and **a** is the minor radius. If however, **a > b > 0**, it is horizontal, **a** is the major, and **b** is the minor.

Next, we can notice that the two points (-1, 1) to (-3, 1) share the same *y*-coordinate, so we can expect that the two points are the two endpoints of the axis of symmetry parallel to the *x*-axis. And the axis of symmetry is one of the two main axes.

And also, we can notice that the other two points (-2, -1) and (-2, 3) share the same *x*-coordinate, so we can also expect that the two points are the two endpoints of the axis of symmetry parallel to the *y*-axis.

How then, can we check to see if the four points are the four endpoints of the main axes?

We know that the center is the midpoint of the vertices, which are the two endpoints of the major axis, and also, that the center is the midpoint of the two endpoints of the minor axis, too.

So we can check to see of the midpoint of the two points (-1, 1) to (-3, 1) is the same as the midpoint of the other two points (-2, -1) and (-2, 3).

Taking thus, the midpoint of (-1, 1) to (-3, 1), we get: (-2, 1). And taking the midpoint of (-2, -1) and (-2, 3), we get: (-2, 1), too. So the four points are the four endpoints of the main axes, and the center is (-2, 1).

So next, taking the distance between (-1, 1) to (-3, 1), we get: 2. And taking the distance between (-2, -1) and (-2, 3), we get: 4. So the major axis is 4, and the minor axis is 2.

And the vertices are the endpoints of the major axis, so two points (-2, -1) and (-2, 3) are the two vertices. And if the ellipse is vertical, the vertices share the same x-coordinate. So the ellipse we want to find is vertical.

So the ellipse is: $(x+2)^2 + \dfrac{(y-1)^2}{4} = 1$, often put this way, too: $\dfrac{(x+2)^2}{1^2} + \dfrac{(y-1)^2}{2^2} = 1.$

Fig. 0

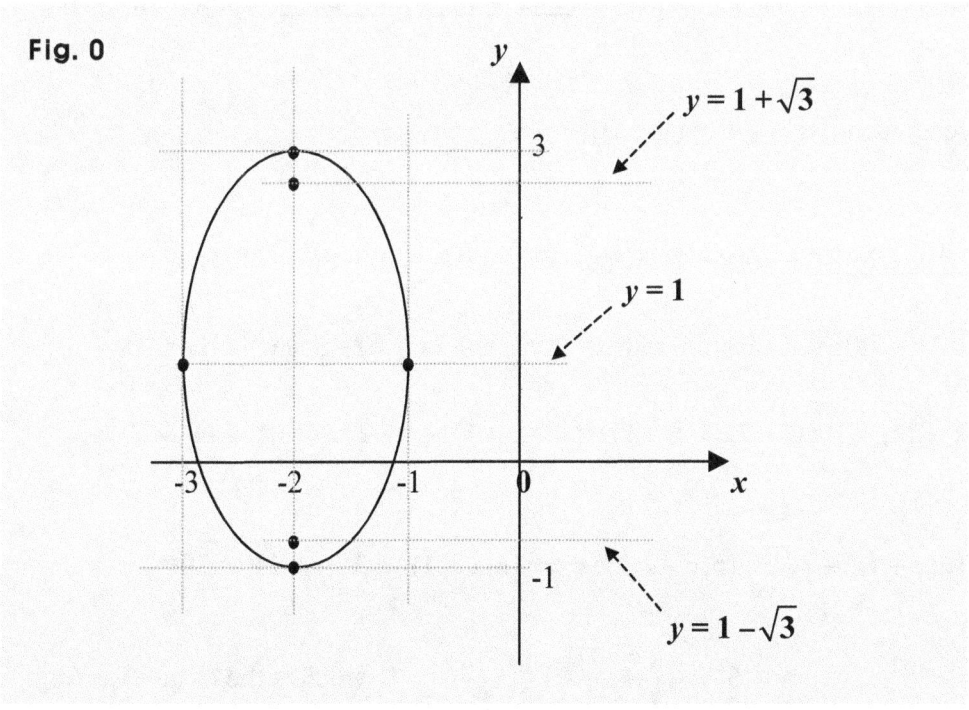

Suggestions or Solutions
To the **Problem** in the Example **3**

Find the ellipse of which the eccentricity is 3/5, and a directrix is a line $x = 28/3$, and is corresponding to a focus at (4, 2).

Suppose first, $T(x, y)$ is an arbitrary point of an ellipse, D is the distance from T to a directrix, F is the distance from T to the focus corresponding to the directrix, and e is the eccentricity. Then, we get: $F = eD$, that is, $e = F/D$. So in this case, we get: $F/D = 3/5$.

So next, taking the distances, we get: $D^2 = (x - 28/3)^2$, and $F^2 = (x - 4)^2 + (y - 2)^2$.

Then, we get: $F/D = 3/5 \Rightarrow (F/D)^2 = F^2/D^2 = 9/25 = \dfrac{(x-4)^2 + (y-2)^2}{(x - \frac{28}{3})^2}$

$\Rightarrow 9(x - 28/3)^2 = 25\{(x - 4)^2 + (y - 2)^2\} \Rightarrow \underline{25(x - 4)^2 - 9(x - 28/3)^2} + 25(y - 2)^2 = 0$.

And we have an identity: $A - B = (A + B)(A - B)$.   So we get:

$\underline{25(x - 4)^2 - 9(x - 28/3)^2} = \{5(x - 4) + 3(x - 28/3)\}\{5(x - 4) - 3(x - 28/3)\}$

$= (5x - 20 + 3x - 28)(5x - 20 - 3x + 28) = (8x - 48)(2x + 8) = 16(x - 6)(x + 4)$

$= 16(x^2 - 2x - 24) = 16(x^2 - 2x + 1 - 25) = 16(x - 1)^2 - 16 \cdot 25 = 16(x - 1)^2 - 20^2$.

So we get: $16(x - 1)^2 - 20^2 + 25(y - 2)^2 = 0 \Rightarrow 16(x - 1)^2 + 25(y - 2)^2 = 20^2$

$\Rightarrow 4^2(x - 1)^2 + 5^2(y - 2)^2 = 4^2 5^2 \Rightarrow \dfrac{(x-1)^2}{5^2} + \dfrac{(y-2)^2}{4^2} = 1$, which is the ellipse we want.

Suggestions or Solutions
To the **Problem** in the Example **4**

Suppose C is a point in a line segment $\overline{AB}$, and $\overline{AC}/\overline{CB} = 1/2$. Then, assuming the end point A is moving along the x-axis, and at the same time, the end point B is moving along the y-axis, find the curve that the point C makes.

Setting first, $C = (x, y)$, $A = (u, 0)$, $B = (0, v)$, and $\overline{AB} = k$, we get: $x = 2u/3$ and $y = v/3$.

Then, by the distance formula, we get: $(\overline{AB})^2 = (u - 0)^2 + (0 - v)^2 = k^2 \Rightarrow u^2 + v^2 = k^2$. And we get: $x = 2u/3 \Rightarrow u = 3x/2$, and $y = v/3 \Rightarrow v = 3y$.

So next, we get: $u^2 + v^2 = k^2 \Rightarrow (3x/2)^2 + (3y)^2 = \dfrac{x^2}{\left(\frac{2}{3}\right)^2} + \dfrac{y^2}{\left(\frac{1}{3}\right)^2} = k^2 \Rightarrow \dfrac{x^2}{\left(\frac{2}{3}k\right)^2} + \dfrac{y^2}{\left(\frac{1}{3}k\right)^2} = 1$,

which is a horizontal ellipse where the center is $(0, 0)$, the major radius is $2k/3$, and the minor radius is $k/3$.

If not quite sure of the idea behind the processes above, follow the steps below:

Setting first, $A = (u, 0)$, and $B = (0, v)$, we can put the point C and the line segment $\overline{AB}$ in a graph the way below:

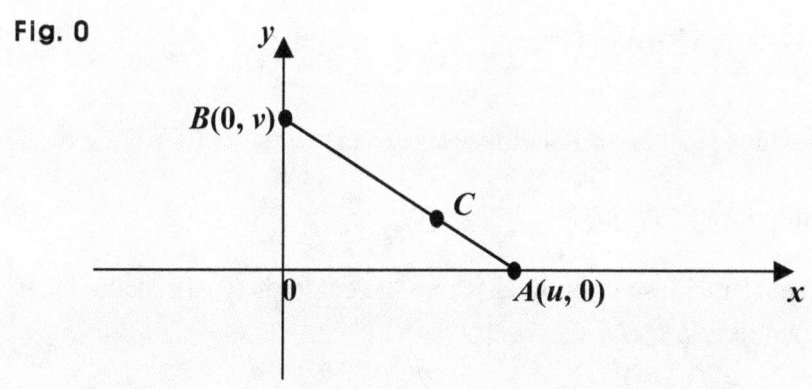

Fig. 0

Next, we know that the point C moves in the x-y plane as the two points A and B, and makes a curve. So we can take the point C as the arbitrary point in the curve, and thus, can set: $C = (x, y)$.

Next, assuming: $\overline{AB} = k$, that is, the length of $\overline{AB}$ is k, by the distance formula, we get: $(\overline{AB})^2 = (u-0)^2 + (0-v)^2 = k^2 \Rightarrow u^2 + v^2 = k^2$.

Next, we know: $\overline{AC} / \overline{CB} = 1/2$. So we get: $x = 2u/3$ and $y = v/3$.
In other words, the point C is at the third of the line segment from the point A. It's because $\overline{CB}$ is twice $\overline{AC}$.

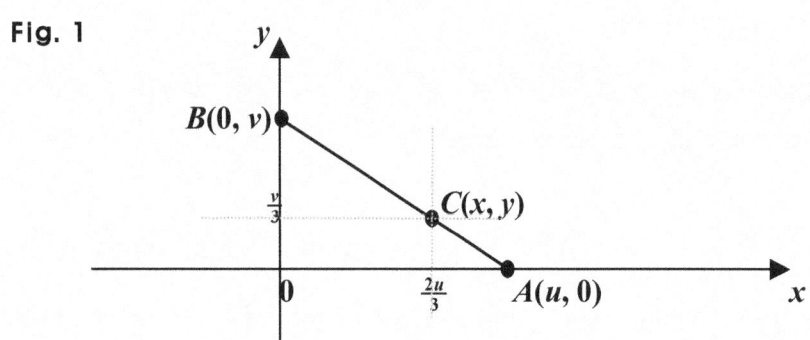

Fig. 1

Then, we get: $x = 2u/3 \Rightarrow u = 3x/2$, and $y = v/3 \Rightarrow v = 3y$.

So next, getting the connective equation between x and y, we get the equation of the curve, and getting the equation, we get:

$$u^2 + v^2 = k^2 \Rightarrow (\tfrac{3}{2}x)^2 + (3y)^2 = k^2 \Rightarrow \left(\frac{x}{\tfrac{2}{3}}\right)^2 + \left(\frac{y}{\tfrac{1}{3}}\right)^2 = k^2 \Rightarrow \frac{x^2}{(\tfrac{2}{3})^2} + \frac{y^2}{(\tfrac{1}{3})^2} = k^2$$

$$\Rightarrow \frac{x^2}{(\tfrac{2}{3}k)^2} + \frac{y^2}{(\tfrac{1}{3}k)^2} = 1, \quad \text{which is a horizontal ellipse where the center is } (0, 0), \text{ the major}$$

radius is $2k/3$, and the minor radius is $k/3$.

In other words, the curve is an ellipse horizontal where the center is $(0, 0)$, the major axis is $4k/3$, and the minor axis is $2k/3$.

What then, about the case where $\overline{BC} / \overline{CA} = 1/2$?

Setting again, $A = (u, 0)$, and $B = (0, v)$, we can put the point C and the line segment $\overline{AB}$ in a graph the way below:

Fig. 2

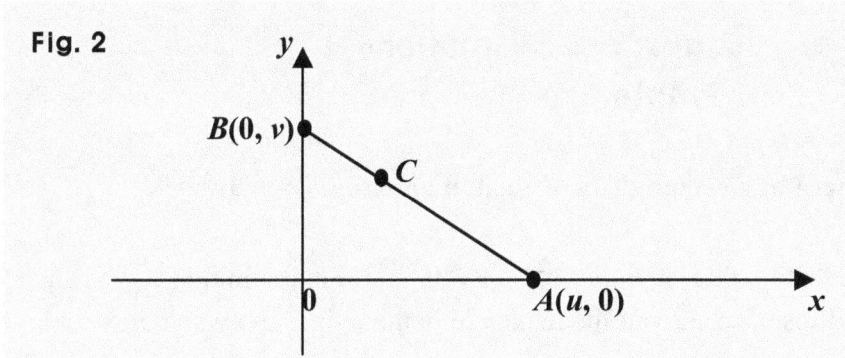

And taking the point C as the arbitrary point in the curve, we can set: $C = (x, y)$.
Next, assuming again: $\overline{AB} = k$, we get: $(\overline{AB})^2 = (u - 0)^2 + (0 - v)^2 = k^2 \Rightarrow u^2 + v^2 = k^2$.

Next, we know: $\overline{BC} / \overline{CA} = 1 / 2$. So we get: $x = 2u/3$ and $y = v/3$.

In other words, the point C is at the third of the line segment from the point A. It's
because $\overline{CB}$ is twice $\overline{AC}$.

Fig. 2

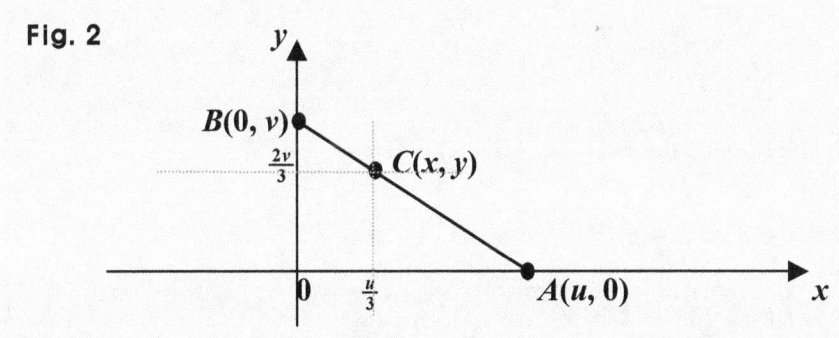

Then, we get: $x = u/3 \Rightarrow u = 3x$, and $y = 2v/3 \Rightarrow v = 3y/2$.

So next, getting the connective equation between x and y, we get:

$$u^2 + v^2 = k^2 \Rightarrow (3x)^2 + (\tfrac{3}{2}y)^2 = k^2 \Rightarrow \left(\frac{x}{\frac{1}{3}}\right)^2 + \left(\frac{y}{\frac{2}{3}}\right)^2 = k^2 \Rightarrow \frac{x^2}{(\frac{1}{3})^2} + \frac{y^2}{(\frac{2}{3})^2} = k^2$$

$$\Rightarrow \frac{x^2}{(\frac{1}{3}k)^2} + \frac{y^2}{(\frac{2}{3}k)^2} = 1,$$ which is a vertical ellipse where the center is $(0, 0)$, the major axis

is **4k/3**, so the major radius is **2k/3**, and the minor axis is **2k/3**, so the minor radius is **k/3**.

Suggestions or Solutions
To the **Problem** in the Example **5**

Find the maximum area of a rectangle inscribed in an ellipse $x^2 + 9y^2 = 9$.

First, we can put the equation this way, too: $x^2/3^2 + y^2/1^2 = 1$. So assuming (x, y) is an arbitrary point in the ellipse, we can put the rectangle in the ellipse the way below:

Fig. 0

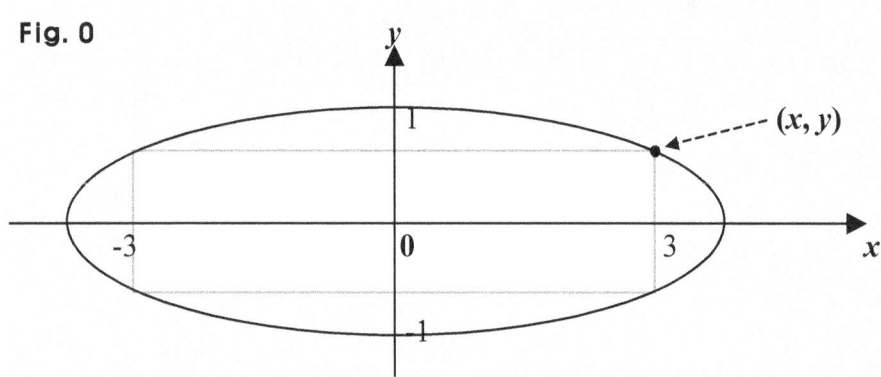

So next, assuming A is the area of the rectangle, we can set: $A = 4|xy|$.

Next, we can get: $x^2 + 9y^2 = 9 \Rightarrow x = \pm\sqrt{9 - 9y^2} = \pm3\sqrt{1 - y^2}$.

So we get: $A = 4|xy| = 12\,|\,y\sqrt{1 - y^2}\,| = 12\sqrt{y^2 - y^4}$.

Next, setting: $s = y^2$, we get: $-1 \le y \le 1 \Rightarrow 0 \le y^2 \le 1 \Rightarrow 0 \le s \le 1$.

So next, setting: $t = y^2 - y^4$, we get: $t = s - s^2$ for $0 \le s \le 1$.

And we get: $t = -(s^2 - s) = -(s^2 - s + 1/4 - 1/4) = -(s - 1/2)^2 + 1/4$.

So when $s = 1/2$, t gets its maximum value, which is $1/4$.

That is, when $y^2 = 1/2$, $(y^2 - y^4)$ gets its maximum value, which is $1/4$.

So when $y = \pm\frac{1}{\sqrt{2}} = \pm\frac{\sqrt{2}}{2}$, $A = 12\sqrt{t}$ gets its maximum value, which is: $12\sqrt{\frac{1}{4}} = 6$.

Examples 6 in Ellipses

0.   In the x-y plane, find the area where the points satisfy the inequality as follows:
$(x^2 + y^2 - 9)(x^2 + 4y^2 - 16) \le 0$.

1.   Assuming $9x^2 + ax + y^2 + by + c = 0$ is an ellipse, find the values of a, b, and c that make the ellipse tangent to the x-axis at a point $(1, 0)$ and pass through a point $(3, 3)$.

Suggestions or Solutions
To the Problem in the Example 0

In the *x-y* plane, find the area where the points satisfy the inequality as follows:
$(x^2 + y^2 - 9)(x^2 + 4y^2 - 16) \leq 0.$

In the graph below, the area we want is the area with slash marks, and the area includes the circle and ellipse themselves, too.

Fig. 0

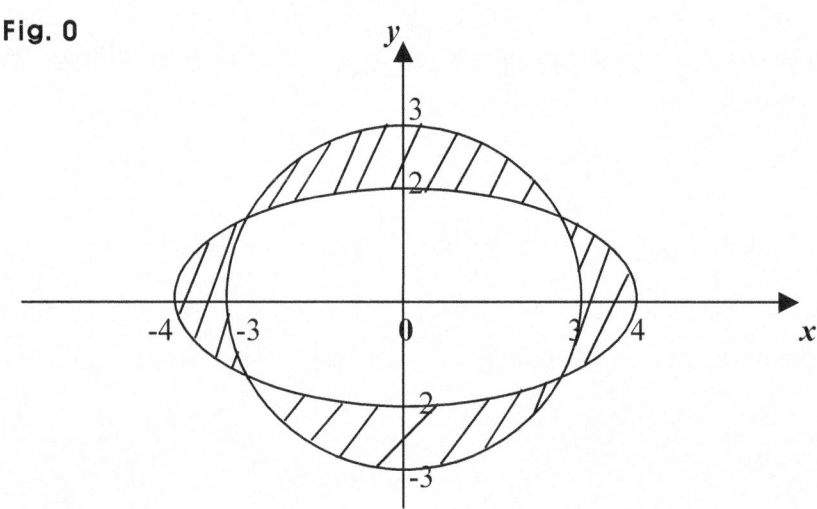

If not quite sure of the idea behind the processes above, follow the steps below:

Assuming first, $AB \leq 0$, we get two cases.
One is: $A \leq 0$ and $B \geq 0$, and the other is: $A \geq 0$ and $B \leq 0$.

So if $(x^2 + y^2 - 9)(x^2 + 4y^2 - 16) \leq 0$, we get two cases, too.

One is: $x^2 + y^2 - 9 \leq 0$ and $x^2 + 4y^2 - 16 \geq 0$.
And the other is: $x^2 + y^2 - 9 \geq 0$, and $x^2 + 4y^2 - 16 \leq 0$.

And we have: $x^2 + y^2 \leq 9 = 3^2$, and $x^2 + 4y^2 \geq 16 \Rightarrow x^2/16 + y^2/4 \geq 1 \Rightarrow x^2/4^2 + y^2/2^2 \geq 1$.

So in one case, we have: $x^2 + y^2 \leq 3^2$, and $x^2/4^2 + y^2/2^2 \geq 1$.

And in the other case, we have: $x^2 + y^2 \geq 3^2$, and $x^2/4^2 + y^2/2^2 \leq 1$.

Next, $x^2 + y^2 = 9 = 3^2$ is the equation of a circle centered at the origin with a radius of 3.

And $x^2/4^2 + y^2/2^2 = 1$ is the equation of a horizontal ellipse centered at the origin with a semi major axis of 4 and a semi minor axis of 2.

Next, if $(x - u)^2 + (y - v)^2 \geq r^2$, the inequality means the set of all the points in the area <u>outside</u> a circle centered at (u, v) with a radius of r, and also, the area includes the circle itself, too, because of the equal sign.

Fig. 1

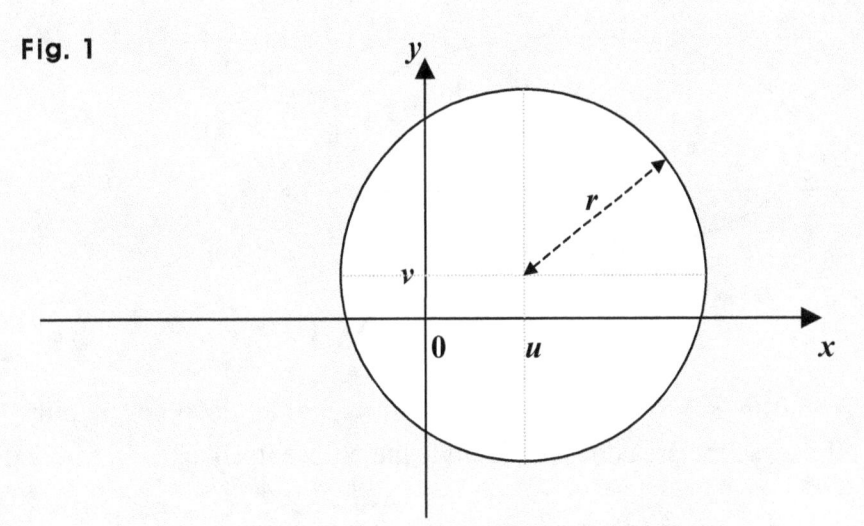

So if $(x - u)^2 + (y - v)^2 < r^2$, the inequality means the set of all the points in only the area inside the circle, so the area does not include the circle itself.

Fig. 2

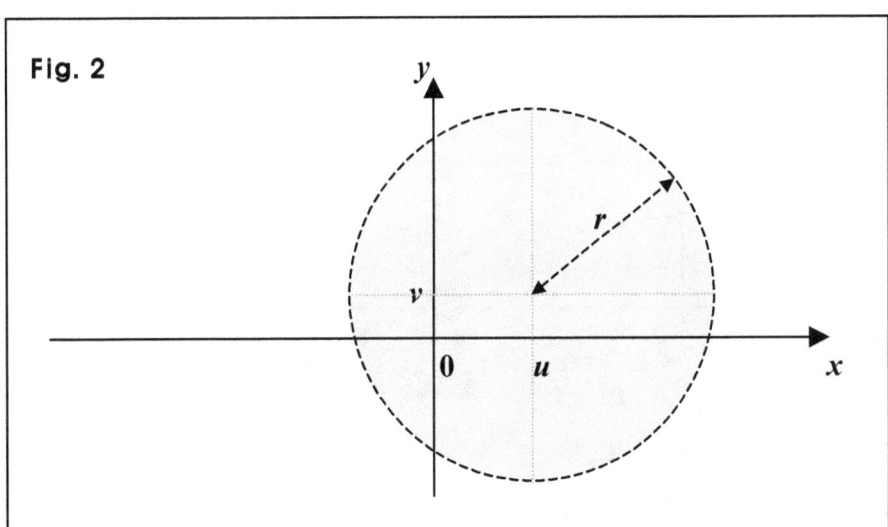

And the same is true for an ellipse, too.

So if $(x - u)^2/a^2 + (y - v)^2/b^2 \geq 1$, the inequality means the set of all the points in the area outside an ellipse centered at (u, v) with the main axes of **2a** and **2b**, and also, the area includes the ellipse itself, too. And if $a > b > 0$, we get:

Fig. 0

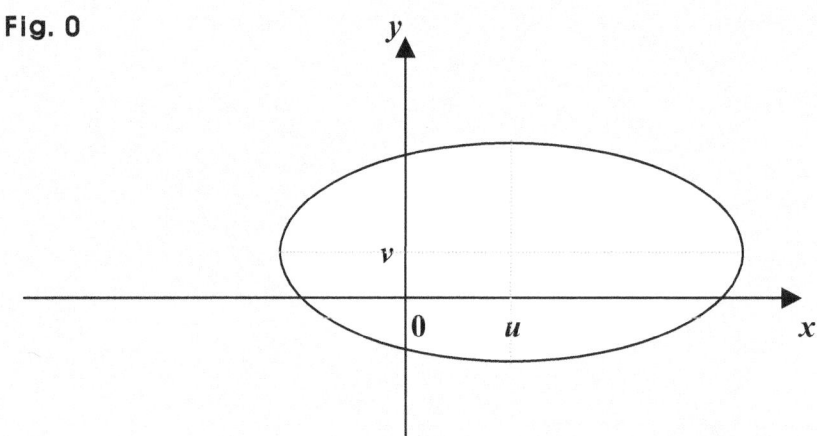

And if $(x - u)^2/a^2 + (y - v)^2/b^2 < 1$, the inequality means the set of all the points in only the area insides the ellipse, so the area does not include the ellipse itself.

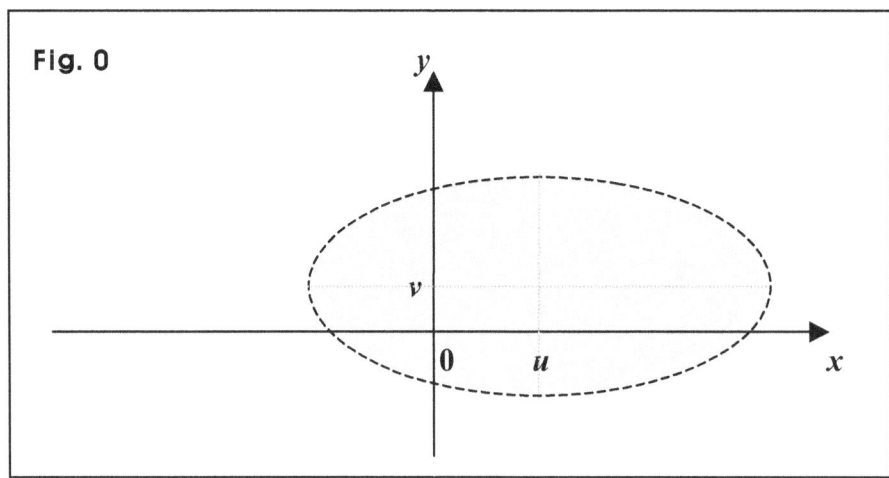

Now, we have two cases.

One is: $x^2 + y^2 \leq 3^2$, and $x^2/4^2 + y^2/2^2 \geq 1$.

And the other is: $x^2 + y^2 \geq 3^2$, and $x^2/4^2 + y^2/2^2 \leq 1$.

So in the graph below, the area we want is the area with slash marks, and the area includes the circle and ellipse themselves, too.

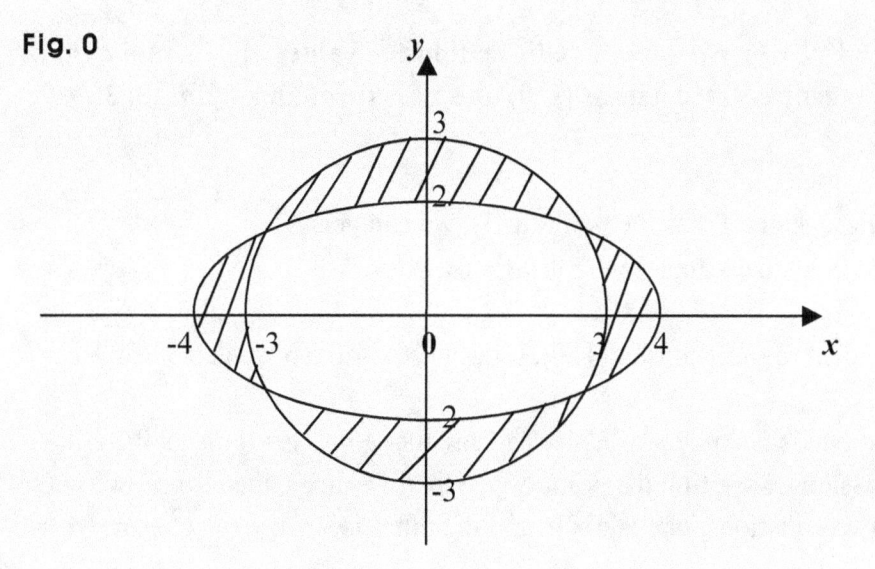

Fig. 0

What if we have: $(x^2 + y^2 - 9)(x^2 + 4y^2 - 16) > 0$?

Then, we get two cases, too, but the area is the opposite of the area with slash marks above.

So one is: $x^2 + y^2 > 3^2$, and $x^2/4^2 + y^2/2^2 > 1$.

And the other is: $x^2 + y^2 < 3^2$, and $x^2/4^2 + y^2/2^2 < 1$.

So in the graph above, the area is the area without slash marks, and the area does not include the circle and ellipse themselves.

Suggestions or Solutions
To the Problem in the Example 1

Assuming $9x^2 + ax + y^2 + by + c = 0$ is an ellipse, find the values of a, b, and c that make the ellipse tangent to the x-axis at $(1, 0)$ and pass through a point $(3, 3)$.

First, if E is the ellipse, since E has the point $(3, 3)$, we can get:
$9(3^2) + 3a + 3^2 + 3b + c = 0 \Rightarrow 3a + 3b + c + 90 = 0$.

Next, since E is tangent to the x-axis at $(1, 0)$, it has $(1, 0)$, too, so we get: $\mathbf{9 + a + c = 0}$.

And next, since the x-axis is tangent to E, we can say that a line $y = 0$ meets the ellipse E at one point. And assuming we find the point where the line meets the ellipse, we get to solve a system of two equations, one is $y = 0$, and the other is $9x^2 + ax + y^2 + by + c = 0$.

And solving the system, since $y = 0$, we can get first, $9x^2 + ax + 0^2 + b \cdot 0 + c = 0$, which is: $9x^2 + ax + c = 0$, which is a quadratic equation, which therefore, has to have a double root, because the line meets the ellipse at one point.

So the discriminant of the quadratic equation has to be 0.
Thus, we get $a^2 - 4 \cdot 9c = a^2 - 36c = 0$.

So we get a system of three equations as follows:
$3a + 3b + c + 90 = 0$, $9 + a + c = 0$, and $a^2 - 36c = 0$.

So solving the system, we can get first: $9 + a + c = 0 \Rightarrow c = -a - 9$.

So next, we can get: $a^2 - 36c = 0 \Rightarrow a^2 - 36(-a - 9) = 0 \Rightarrow a^2 + 36a + 36 \cdot 9 = 0$.

Meanwhile, $36 \cdot 9 = 4 \cdot 9 \cdot 9 = 2 \cdot 9 \cdot 2 \cdot 9 = 18^2$, and $36 = 2 \cdot 18$.

So we get: $a^2 + 36a + 36 \cdot 9 = a^2 + 2 \cdot 18a + 18^2 = (a + 18)^2 = 0 \Rightarrow a = -18$.

Thus next, we can get: $c = -a - 9 = 18 - 9 = 9 \Rightarrow c = 9$.
And next, $3a + 3b + c + 90 = 0 \Rightarrow 3b = -3a - c - 90 = 54 - 9 - 90 = -45 \Rightarrow b = -15$.

What ellipse then, is it?

There can be infinitely many ellipses that can be tangent to the *x*-axis at (1, 0), and pass through (3, 3).

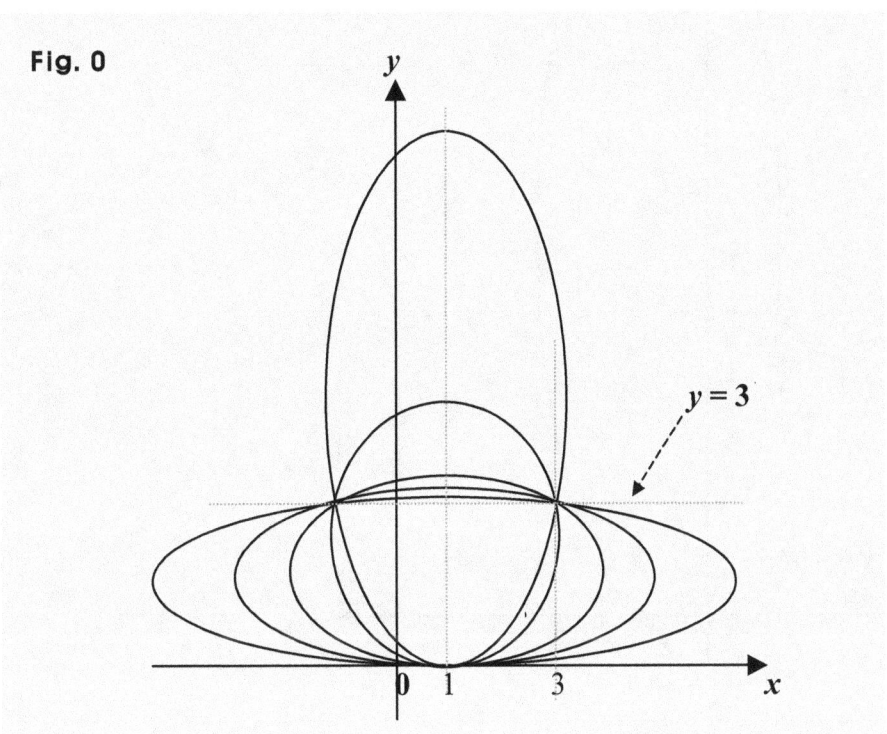

Fig. 0

$y = 3$

Next, putting values of *a*, *b*, and *c* into the equation $9x^2 + ax + y^2 + by + c = 0$, we get:

$$9x^2 - 18x + y^2 - 15y + 9 = 9(x^2 - 2x + 1 - 1) + y^2 - 15y + (\tfrac{15}{2})^2 - (\tfrac{15}{2})^2 + 9$$

$$= 9(x-1)^2 - 9 + (y - \tfrac{15}{2})^2 - (\tfrac{15}{2})^2 + 9 = 9(x-1)^2 + (y - \tfrac{15}{2})^2 - (\tfrac{15}{2})^2 = 0$$

$$\Rightarrow \frac{9(x-1)^2}{(\tfrac{15}{2})^2} + \frac{(y - \tfrac{15}{2})^2}{(\tfrac{15}{2})^2} - 1 = 0 \Rightarrow \frac{(x-1)^2}{(\tfrac{5}{2})^2} + \frac{(y - \tfrac{15}{2})^2}{(\tfrac{15}{2})^2} = 0.$$

So the ellipse *E* is a vertical ellipse centered at $(1, \tfrac{15}{2})$ with the main axes of 5 and 15.

Fig. 1

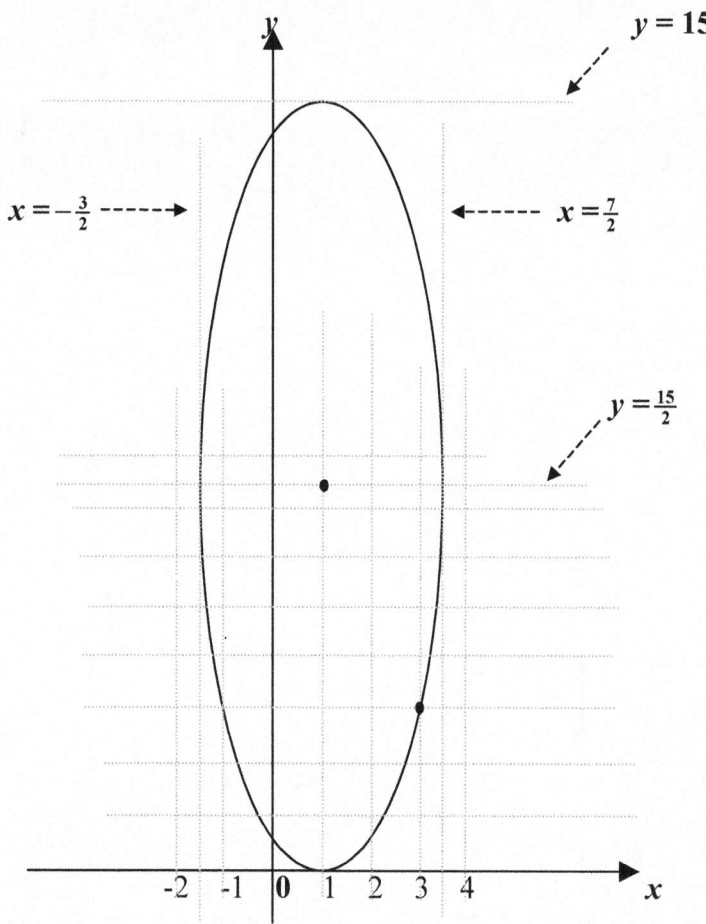

$y = 15$

$x = -\frac{3}{2}$

$x = \frac{7}{2}$

$y = \frac{15}{2}$

Examples 7 in Ellipses

Find the equation of the line tangent to an ellipse $x^2/a^2 + y^2/b^2 = 1$ at a point (s, t).

Suggestions or Solutions
To the Problem in the Example

Find the equation of the line tangent to an ellipse $x^2/a^2 + y^2/b^2 = 1$ at a point (s, t).

Assuming first, the tangent line is $y = mx + n$, we get first, $x^2/a^2 + (mx + n)^2/b^2 = 1$. And rearranging terms in the equation above, we get:

$$x^2/a^2 + (mx + n)^2/b^2 = 1 \Rightarrow b^2x^2 + a^2(mx + n)^2 = a^2b^2$$

$$\Rightarrow b^2x^2 + a^2(mx + n)^2 = b^2x^2 + a^2(m^2x^2 + 2mnx + n^2)$$

$$= (b^2 + a^2m^2)x^2 + 2mna^2x + a^2n^2 = a^2b^2 \Rightarrow (b^2 + a^2m^2)x^2 + 2mna^2x + a^2(n^2 - b^2) = 0.$$

Next, since the line meets the ellipse at one point, the equation's discriminant has to be 0.

So we get: $m^2n^2a^4 - (b^2 + a^2m^2)a^2(n^2 - b^2) = 0 \Rightarrow m^2n^2a^2 - (b^2 + a^2m^2)(n^2 - b^2) = 0$

$$\Rightarrow b^4 - b^2n^2 + a^2b^2m^2 = 0 \Rightarrow b^2 - n^2 + a^2m^2 = 0.$$

Next, we know the fact that the line passes through (s, t).   So we get: $t = ms + n$. Thus, we get: $n = t - ms$.   So we get:

$$b^2 - n^2 + a^2m^2 = b^2 - (t - ms)^2 + a^2m^2 = 0$$

$$\Rightarrow b^2 - t^2 + 2stm - s^2m^2 + a^2m^2 = m^2(a^2 - s^2) + 2stm + b^2 - t^2 = 0.$$

So we get: $m = \dfrac{-st \pm \sqrt{s^2t^2 - (a^2 - s^2)(b^2 - t^2)}}{a^2 - s^2}$. We know however, m can have one

value only.   So what's inside the square root has to be 0.   That is to say that the discriminant is 0.

So we get: $m = \dfrac{-st}{a^2 - s^2}$. And since the discriminant is 0, we get:

$$s^2t^2 - (a^2 - s^2)(b^2 - t^2) = s^2b^2 + a^2t^2 - a^2b^2 = 0 \Rightarrow b^2(s^2 - a^2) = -a^2t^2 \Rightarrow a^2 - s^2 = \dfrac{a^2t^2}{b^2}.$$

So we get: $m = -\dfrac{sb^2}{ta^2}$. And we know that the line passes through (s, t).

So the tangent line is: $y - t = -\dfrac{sb^2}{ta^2}(x - s)$. And it can be put this way, too: $\dfrac{sx}{a^2} + \dfrac{ty}{b^2} = 1$.

If not quite sure of the idea behind the processes above, follow the steps below:

Suppose first, the tangent line is $y = mx + n$, and is called T, and the ellipse is E.

Then, since the line T is tangent to the ellipse E, we can say that the line T meets the ellipse E at one point.

And assuming we find the point where the line meets the ellipse, we get to solve a system of two equations, one is $y = mx + n$, and the other is $x^2/a^2 + y^2/b^2 = 1$.

And solving the system, since $y = mx + n$, we can get first, $x^2/a^2 + (mx + n)^2/b^2 = 1$, which is: $b^2x^2 + a^2(mx + n)^2 = a^2b^2$, which is a quadratic equation, which therefore, has to have a double root, because the line meets the ellipse at one point.

So the discriminant of the quadratic equation has to be 0.
By the way, if the coefficient of x is even, we can take a quarter of the discriminant, because the result will get simplified that way.

If for instance, we have: $Ax^2 + 2Bx + C = 0$, the discriminant is: $B^2 - AC$.

So first, rearranging terms in the equation, we get:

$$b^2x^2 + a^2(mx + n)^2 = b^2x^2 + a^2(m^2x^2 + 2mnx + n^2)$$

$$= (b^2 + a^2m^2)x^2 + 2mna^2x + a^2n^2 = a^2b^2 \Rightarrow (b^2 + a^2m^2)x^2 + 2mna^2x + a^2(n^2 - b^2) = 0.$$

So next, getting the quarter of the discriminant, we get:

$$m^2n^2a^4 - (b^2 + a^2m^2)a^2(n^2 - b^2) = 0 \Rightarrow m^2n^2a^2 - (b^2 + a^2m^2)(n^2 - b^2) = 0$$
$$\Rightarrow m^2n^2a^2 - b^2n^2 + b^4 - m^2n^2a^2 + a^2b^2m^2 = b^4 - b^2n^2 + a^2b^2m^2 = 0$$
$$\Rightarrow b^2 - n^2 + a^2m^2 = 0.$$

Next, knowing the slope and a point of a line, we can get the equation of the line.

We know the tangent line T passes through (s, t), and the slope is m.
And also, s, t, a, and b are all known values.

So finding m, we can get the line T.   How then, can we find m?

We now have two equations that have **m**.

One is: $b^2 - n^2 + a^2m^2 = 0$. and the other is: $y = mx + n$.

And we know the fact that **T** passes through **(s, t)**.   So we get: $t = ms + n$.
Thus, we get: $n = t - ms$.

So we can now get **m** putting **n** into the other equation.   Then, we get:

$b^2 - n^2 + a^2m^2 = b^2 - (t - ms)^2 + a^2m^2 = 0$
$\Rightarrow b^2 - t^2 + 2stm - s^2m^2 + a^2m^2 = m^2(a^2 - s^2) + 2stm + b^2 - t^2 = 0$, which is quadratic.

So using the quadratic formula, we get: $m = \dfrac{-st \pm \sqrt{s^2t^2 - (a^2 - s^2)(b^2 - t^2)}}{a^2 - s^2}$.

We know however, **m** can have one value only.

So what's inside the square root has to be 0. That is to say that the discriminant is 0.

So we get: $m = \dfrac{-st}{a^2 - s^2} = \dfrac{st}{s^2 - a^2}$.

And since the discriminant is 0, we get:

$s^2t^2 - (a^2 - s^2)(b^2 - t^2) = s^2t^2 - a^2b^2 + a^2t^2 + s^2b^2 - s^2t^2 = s^2b^2 + a^2t^2 - a^2b^2 = 0$

$\Rightarrow s^2b^2 - a^2b^2 + a^2t^2 = 0 \Rightarrow b^2(s^2 - a^2) = -a^2t^2 \Rightarrow s^2 - a^2 = \dfrac{-a^2t^2}{b^2}$.

So we get: $m = \dfrac{st}{s^2 - a^2} = st\dfrac{b^2}{-a^2t^2} = -\dfrac{sb^2}{ta^2}$, which is the slope of the line **T**.

And we know that the line **T** passes through **(s, t)**.
If a line has a slope of **k**, and has a point **(p, q)**, the line is: $y - q = k(x - p)$.

So the line **T** is: $y - t = -\dfrac{sb^2}{ta^2}(x - s)$, which looks a bit complicated.

So simplifying the equation above, we can get:

$$y - t = -\frac{sb^2}{ta^2}(x - s) \Rightarrow ta^2(y - t) = -sb^2(x - s) \Rightarrow tya^2 - t^2a^2 = -sxb^2 + s^2b^2$$

$$\Rightarrow sxb^2 + tya^2 = s^2b^2 + t^2a^2 \Rightarrow sx + \frac{tya^2}{b^2} = s^2 + \frac{t^2a^2}{b^2} \Rightarrow \frac{sx}{a^2} + \frac{ty}{b^2} = \frac{s^2}{a^2} + \frac{t^2}{b^2}.$$

And we know that the ellipse E passes through the point (s, t), and E is: $\frac{x^2}{a^2} + \frac{y^2}{b^2} = 1$.

So we get: $\frac{s^2}{a^2} + \frac{t^2}{b^2} = 1$, and thus, the tangent line is: $\frac{sx}{a^2} + \frac{ty}{b^2} = 1$.

So for instance, finding the line tangent to $\frac{x^2}{3^2} + \frac{y^2}{2^2} = 1$ at $(1, \frac{4\sqrt{2}}{3})$, we get:

$$\frac{x}{3^2} + \frac{4\sqrt{2}y}{3 \cdot 2^2} = \frac{x}{9} + \frac{\sqrt{2}y}{3} = 1, \text{which is the tangent line.}$$

And of course, we can put it this way, too: $x + 3\sqrt{2}y - 9 = 0$.

What if we are given the slope of the tangent line and are asked to find the tangent line?

Assuming the slope given is m, we can assume that the tangent line is $y = mx + n$.

Then, finding n, we get the line tangent to the ellipse.

Finding n, we just solve $b^2 - n^2 + a^2m^2 = 0$ for n. And we get the equation taking the discriminant of $x^2/a^2 + (mx + n)^2/b^2 = 1$, which is: $b^2x^2 + a^2(mx + n)^2 = a^2b^2$.

So solving it for n, we get: $n = \pm\sqrt{b^2 + a^2m^2}$. And thus, we get two tangent lines.

One is: $y = mx + \sqrt{b^2 + a^2m^2}$, and the other is: $y = mx - \sqrt{b^2 + a^2m^2}$.

Fig. 0

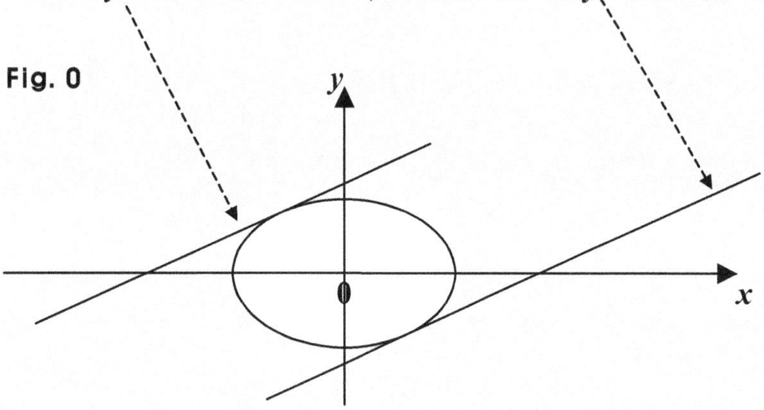

So for instance, finding the lines that have a slope of 2, and are tangent to $\dfrac{x^2}{3^2} + \dfrac{y^2}{2^2} = 1$,

we get: $y = 2x \pm \sqrt{2^2 + 3^2 2^2}$. So we get two tangent lines.

One is $y = 2x + 2\sqrt{10}$, and the other is: $y = 2x - 2\sqrt{10}$.

What if this time, we want to find the line tangent to an ellipse

$\dfrac{(x-u)^2}{a^2} + \dfrac{(y-v)^2}{b^2} = 1$ at a point (s, t)?

Translating the ellipse $\dfrac{x^2}{a^2} + \dfrac{y^2}{b^2} = 1$ in the amount of u along the x-axis, and in the

amount of v along the y-axis, we get the ellipse $\dfrac{(x-u)^2}{a^2} + \dfrac{(y-v)^2}{b^2} = 1$.

So finding the line tangent to $\dfrac{(x-u)^2}{a^2} + \dfrac{(y-v)^2}{b^2} = 1$ at the point (s, t), we find first the

line tangent to $\dfrac{x^2}{a^2} + \dfrac{y^2}{b^2} = 1$ at $(s - u, t - v)$. Then, we get: $\dfrac{(s-u)x}{a^2} + \dfrac{(t-v)y}{b^2} = 1$.

Next, translating the line above in the amount of **u** along the **x**-axis, and in the amount of **v** along the **y**-axis, we get the line tangent to $\dfrac{(x-u)^2}{a^2}+\dfrac{(y-v)^2}{b^2}=1$ at the point **(s, t)**.

Then, we get: $\dfrac{(s-u)(x-u)}{a^2}+\dfrac{(t-v)(y-v)}{b^2}=1$, which is the line tangent to $\dfrac{(x-u)^2}{a^2}+\dfrac{(y-v)^2}{b^2}=1$ at the point **(s, t)**.

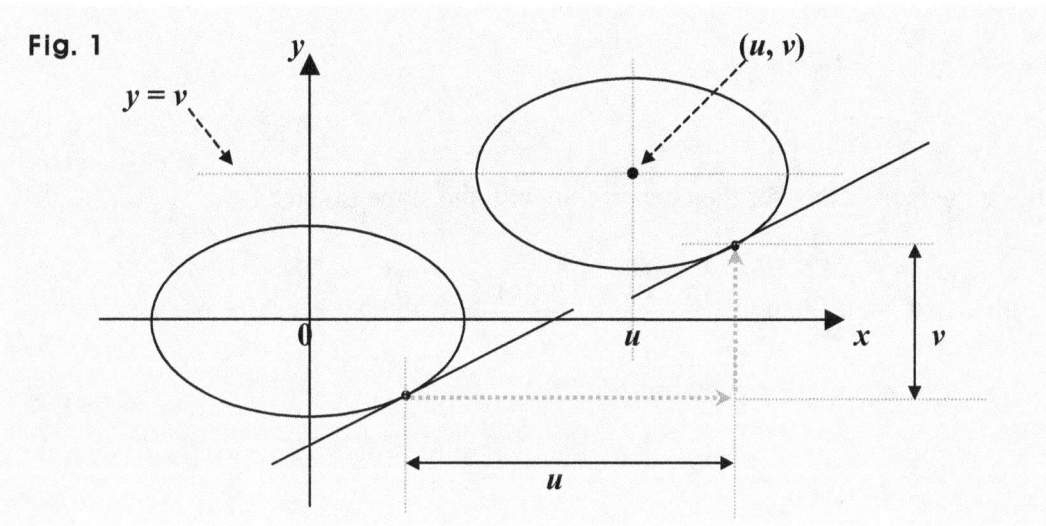

Fig. 1

So for instance, finding the line tangent to $\dfrac{(x-1)^2}{3^2}+\dfrac{(y-2)^2}{2^2}=1$ at $(-1,\tfrac{2\sqrt5}{3}+2)$, we can get first, the line tangent to $\dfrac{x^2}{3^2}+\dfrac{y^2}{2^2}=1$ at $(-1-1,\tfrac{2\sqrt5}{3}+2-2)$, that is, $(-2,\tfrac{2\sqrt5}{3})$.

Then, the line is: $\dfrac{-2x}{3^2}+\dfrac{\tfrac{2\sqrt5}{3}y}{2^2}=1$.

Next, translating the line above in the amount of 1 along the **x**-axis, and in the amount of 2 along the **y**-axis, we get the line tangent to $\dfrac{(x-1)^2}{3^2}+\dfrac{(y-2)^2}{2^2}=1$ at $(-1,\tfrac{2\sqrt5}{3}+2)$.

Then, we get: $\dfrac{-2(x-1)}{3^2}+\dfrac{\frac{2\sqrt{5}}{3}(y-2)}{2^2}=1$, which is $\dfrac{-2(x-1)}{9}+\dfrac{\sqrt{5}(y-2)}{6}=1$, which is

the line tangent to $\dfrac{(x-1)^2}{3^2}+\dfrac{(y-2)^2}{2^2}=1$ at $(-1, \frac{2\sqrt{5}}{3}+2)$.

And of course, we can put the line the way below, too:

$$\dfrac{-2(x-1)}{9}+\dfrac{\sqrt{5}(y-2)}{6}=1 \Rightarrow -4(x-1)+3\sqrt{5}\,(y-2)=18.$$

$$\Rightarrow 4x-3\sqrt{5}\,y+14+6\sqrt{5}=0.$$

And the same is true, too, for the tangent line with the slope given.

So the lines that are tangent to $\dfrac{(x-u)^2}{a^2}+\dfrac{(y-v)^2}{b^2}=1$, and have a slope of *m* are as

follows: $y-v=m(x-u)\pm\sqrt{b^2+a^2m^2}$.

Examples 8 in Ellipses

0. Suppose a point $P(s, t)$ is outside an ellipse $x^2/a^2 + y^2/b^2 = 1$, and two lines Q and R meet at the point P and are tangent to the ellipse. Find the slopes of the two tangent lines.

1. A point $P(s, t)$ is in a circle $x^2 + y^2 = a^2 + b^2$, an ellipse $x^2/a^2 + y^2/b^2 = 1$ is inside the circle, and two lines Q and R meet at the point P and are tangent to the ellipse. Then, show that the two tangent lines Q and R are perpendicular to each other.

2. A point $P(s, t)$ is outside an ellipse $x^2/a^2 + y^2/b^2 = 1$, and two lines meet at the point P, and are tangent to the ellipse at two points Q and R. Then, show that the equation of the line connecting the two tangent points Q and R is: $sx/a^2 + ty/b^2 = 1$. (In this case, the point P is called the pole of the line connecting Q and R, and the line is called the polar of the point P.)

3. Two points $P(s, t)$ and $Q(u, v)$ are outside an ellipse $x^2/a^2 + y^2/b^2 = 1$. Then, show that if Q is in the polar of P, P is in the polar of Q, too.

4. Suppose P is a point in an ellipse, A and B are the two foci, and T is a line tangent to the ellipse at the point P.   Suppose also, m and n are acute angles, m is an angle between the line segment PA and the tangent line T, and n is an angle between the line segment PB and the line T.   Then, show that $m = n$.

Suggestions or Solutions
To the Problem in the Example 0

Suppose a point $P(s, t)$ is outside an ellipse $x^2/a^2 + y^2/b^2 = 1$, and two lines Q and R meet at P and are tangent to the ellipse. Find the slopes of the two tangent lines.

Assuming first, $a > b$, and the ellipse is E, and putting in a graph the two tangent lines Q and R, the point P, and the ellipse E, we can put them the way below:

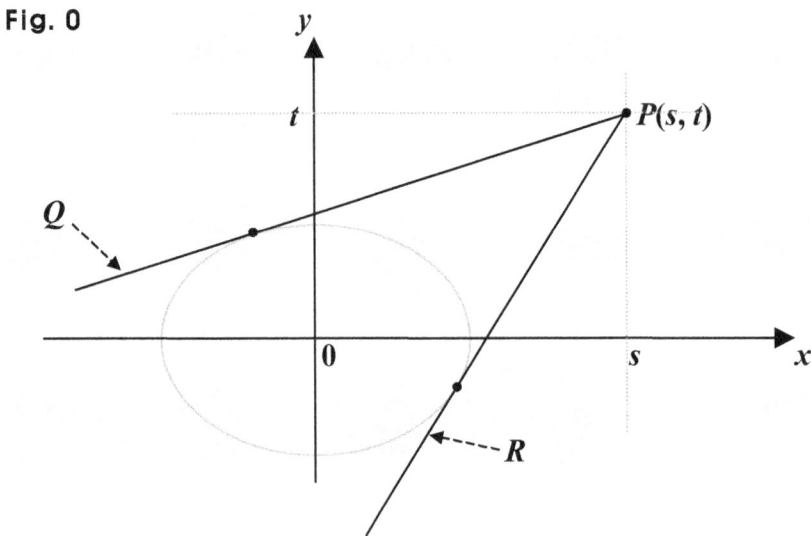

Fig. 0

So examining the graph above, we can notice that this example is quite close to the example in **Examples 7 in Ellipses**.

This is in fact, no other than the example 2, and the only difference is that the point the tangent line passes through is not a tangent point, but is outside the ellipse. So we get two slopes when we get the slope of the tangent line passing through the point $P(s, t)$.

So suppose again, the tangent line Q is: $y = mx + n$.

Then, since the line Q is tangent to the ellipse E, we can say that the line Q meets the ellipse E at one point.

And assuming we find the point where the line meets the ellipse, we get to solve a system of two equations, one is $y = mx + n$, and the other is $x^2/a^2 + y^2/b^2 = 1$.

And solving the system, since $y = mx + n$, we can get first, $x^2/a^2 + (mx + n)^2/b^2 = 1$, which is: $b^2x^2 + a^2(mx + n)^2 = a^2b^2$, which is a quadratic equation, which therefore, has to have a double root, because the line meets the ellipse at one point.

So the discriminant of the quadratic equation has to be 0.

Rearranging terms in the equation first, we get:

$$b^2x^2 + a^2(mx + n)^2 = b^2x^2 + a^2(m^2x^2 + 2mnx + n^2)$$

$$= (b^2 + a^2m^2)x^2 + 2mna^2x + a^2n^2 = a^2b^2 \Rightarrow (b^2 + a^2m^2)x^2 + 2mna^2x + a^2(n^2 - b^2) = 0.$$

Next, getting the quarter of the discriminant, we get:

$$m^2n^2a^4 - (b^2 + a^2m^2)a^2(n^2 - b^2) = 0 \Rightarrow m^2n^2a^2 - (b^2 + a^2m^2)(n^2 - b^2) = 0$$
$$\Rightarrow m^2n^2a^2 - b^2n^2 + b^4 - m^2n^2a^2 + a^2b^2m^2 = b^4 - b^2n^2 + a^2b^2m^2 = 0$$
$$\Rightarrow b^2 - n^2 + a^2m^2 = 0.$$

And we know the fact that Q passes through (s, t), and Q is: $y = mx + n$.
So we get: $t = ms + n$.   Thus, we get: $n = t - ms$.

So we can now get m putting n into the other equation.   Then, we get:

$$b^2 - n^2 + a^2m^2 = b^2 - (t - ms)^2 + a^2m^2 = 0$$
$$\Rightarrow b^2 - t^2 + 2stm - s^2m^2 + a^2m^2 = m^2(a^2 - s^2) + 2stm + b^2 - t^2 = 0, \text{ which is quadratic.}$$

So using the quadratic formula, we get: $m = \dfrac{-st \pm \sqrt{s^2b^2 + a^2t^2 - a^2b^2}}{a^2 - s^2}$.

And thus, we get two slopes. One is the slope of Q, and the other is the slope of R.

One is: $\dfrac{-st + \sqrt{s^2b^2 + a^2t^2 - a^2b^2}}{a^2 - s^2}$. And the other is: $\dfrac{-st - \sqrt{s^2b^2 + a^2t^2 - a^2b^2}}{a^2 - s^2}$.

Suggestions or Solutions
To the Problem in the Example 1

A point $P(s, t)$ is in a circle $x^2 + y^2 = a^2 + b^2$, an ellipse $x^2/a^2 + y^2/b^2 = 1$ is inside the circle, and two lines Q and R meet at the point P and are tangent to the ellipse. Then, show that the two tangent lines Q and R are perpendicular to each other.

Assuming first, $a > b$, and putting in a graph the two tangent lines, the point P, the circle, and the ellipse, we can put them the way below:

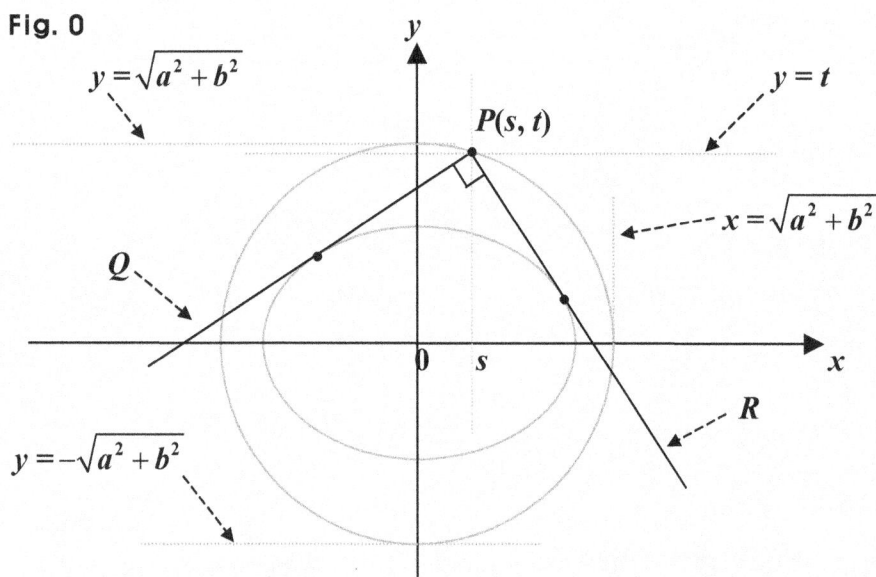

Fig. 0

$y = \sqrt{a^2 + b^2}$

$y = t$

$P(s, t)$

$x = \sqrt{a^2 + b^2}$

Q

$y = -\sqrt{a^2 + b^2}$

R

And we have a fact that if the product of the slopes of two lines is –1, the two lines are perpendicular to each other.   So we can use the fact above.

And in fact, in the example 0 above, we have found the slopes of the two lines Q and R, since the two lines pass through the point $P(s, t)$ outside the ellipse, and are tangent to the ellipse.

And the equation for the two slopes is: $m^2(a^2 - s^2) + 2stm + b^2 - t^2 = 0$.

And we can put it this way: $m^2 + \dfrac{2stm}{a^2 - s^2} + \dfrac{b^2 - t^2}{a^2 - s^2} = 0$.

So showing: $\dfrac{b^2 - t^2}{a^2 - s^2} = -1$, we show that the product of the two slopes is –1. How come?

We have: $(x - u)(y - v) = 0 \Rightarrow x^2 - (u + v)x + uv = 0$. And the two roots are u and v.

So assuming: $ax^2 + bx + c = 0$, we get: $x^2 + (b/a)x + c/a = 0$, and thus, the sum of the two roots is: $-b/a$, and the product of the two roots is c/a.

How then, can we show this: $\dfrac{b^2 - t^2}{a^2 - s^2} = -1$?

We know that the point $P(s, t)$ is in the circle $x^2 + y^2 = a^2 + b^2$.

So we get: $s^2 + t^2 = a^2 + b^2 \Rightarrow b^2 - t^2 = s^2 - a^2 = -(a^2 - s^2) \Rightarrow b^2 - t^2 = -(a^2 - s^2)$.

Thus, we get: $\dfrac{b^2 - t^2}{a^2 - s^2} = -1$.

Let's this time though, actually take the product of the slopes, and see if it is –1.

We know that the slopes are the two roots of the equation as follows:
$m^2(a^2 - s^2) + 2stm + b^2 - t^2 = 0.$

And using the quadratic formula, we get: $m = \dfrac{-st \pm \sqrt{s^2 b^2 + a^2 t^2 - a^2 b^2}}{a^2 - s^2}$.

To begin with, we have an identity: $(A + B)(A - B) = A^2 - B^2$.
So taking the product, we get:

$$\frac{-st + \sqrt{s^2 b^2 + a^2 t^2 - a^2 b^2}}{a^2 - s^2} \cdot \frac{-st - \sqrt{s^2 b^2 + a^2 t^2 - a^2 b^2}}{a^2 - s^2} = \frac{s^2 t^2 - (s^2 b^2 + a^2 t^2 - a^2 b^2)}{(a^2 - s^2)^2}$$

Meanwhile, $s^2 t^2 - (s^2 b^2 + a^2 t^2 - a^2 b^2) = s^2 t^2 - s^2 b^2 - a^2 t^2 + a^2 b^2$
$= a^2(b^2 - t^2) - s^2(b^2 - t^2) = (b^2 - t^2)(a^2 - s^2)$.

So we get: $\dfrac{s^2t^2 - (s^2b^2 + a^2t^2 - a^2b^2)}{(a^2 - s^2)^2} = \dfrac{(b^2 - t^2)(a^2 - s^2)}{(a^2 - s^2)^2} = \dfrac{b^2 - t^2}{a^2 - s^2}$.

And since the point $P(s, t)$ is in the circle $x^2 + y^2 = a^2 + b^2$, we get: $\dfrac{b^2 - t^2}{a^2 - s^2} = -1$.

That's because: $s^2 + t^2 = a^2 + b^2 \Rightarrow b^2 - t^2 = s^2 - a^2 = -(a^2 - s^2) \Rightarrow b^2 - t^2 = -(a^2 - s^2)$.

Consequently, the two tangent lines Q and R are perpendicular to each other.

Suppose now, that U and V are two points where the two tangent lines meet the circle again respectively as shown below.

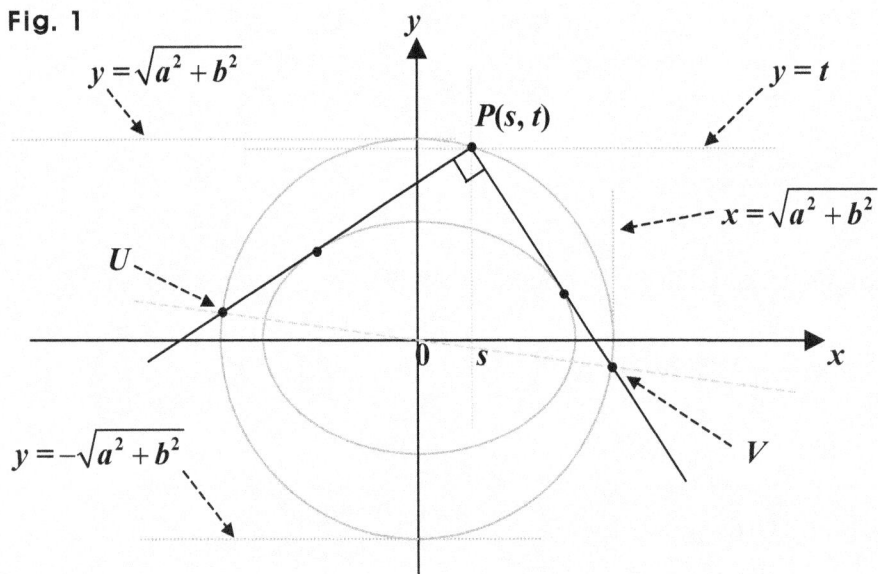

Fig. 1

Then, if the line segment UV passes through the origin, the line segment is the diameter of the circle.

A point in a circle and two endpoints of a line segment that is the diameter of the circle constitute a right triangle.

And since the line segment UV is a diameter and passes through the origin, U and V are symmetric about the center, which is the origin.

So the sum of the x-coordinates at U and V is 0, and so is the sum of the y-coordinates.

Suggestions or Solutions
To the Problem in the Example 2

Assuming a point $P(s, t)$ is outside an ellipse $x^2/a^2 + y^2/b^2 = 1$, and two lines meet at the point P, and are tangent to the ellipse at two points Q and R, show that the equation of the line connecting the two tangent points Q and R is: $sx/a^2 + ty/b^2 = 1$. (In this case, the point P is called the pole of the line connecting Q and R, and the line is called the polar of the point P.)

Assuming first, $a > b$, and putting in a graph the two tangent lines, the points P, Q and R, and the ellipse, we can put them the way below:

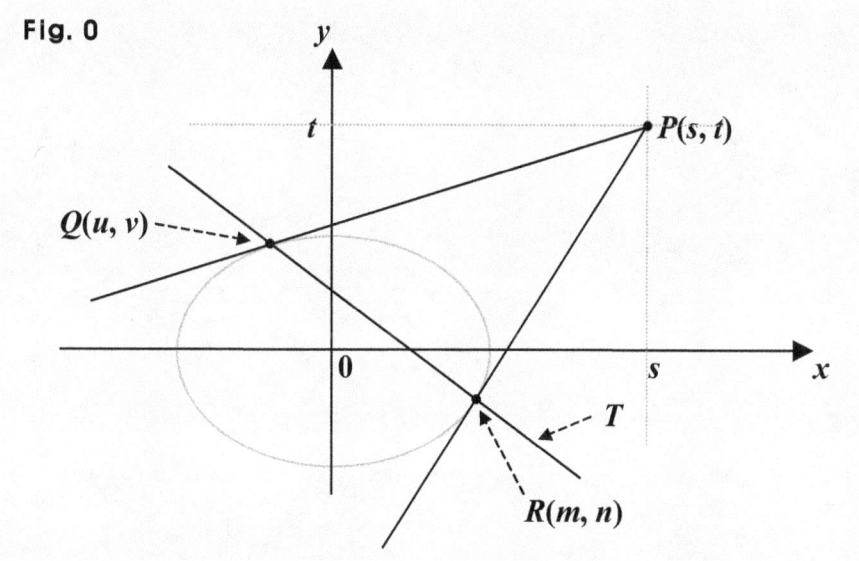

Assuming next, Q is at (u, v), and R is at (m, n), we can say that:

The line tangent to the ellipse at Q is: $ux/a^2 + vy/b^2 = 1$.
The line tangent to the ellipse at R is: $mx/a^2 + ny/b^2 = 1$.

Next, the two lines above pass through the point $P(s, t)$.
So we get: $us/a^2 + vt/b^2 = 1$, and $ms/a^2 + nt/b^2 = 1$.

Assuming thus, T is the line connecting Q and R, we can say that T is: $sx/a^2 + ty/b^2 = 1$, which is the polar of the point P, and passes through the two tangent points Q and R. That's because there is only one line that can pass through two particular points.

Examples 9 in Ellipses

Suppose P is a point in an ellipse, A and B are the two foci, and T is a line tangent to the ellipse at the point P.   Suppose also, m and n are acute angles, m is an angle between the line segment PA and the tangent line T, and n is an angle between the line segment PB and the line T.   Then, show that $m = n$.

174

Suggestions or Solutions
To the Problem in the Example

Suppose *P* is a point in an ellipse, *A* and *B* are the foci, and *T* is a line tangent to the ellipse at *P*.   Suppose also, *m* and *n* are two acute angles, *m* is an angle between *PA* and *T*, and *n* is an angle between *PB* and *T*.   Then, show that *m* = *n*.

As shown below, suppose *Q* is a point outside the ellipse, and *R* is a point moving along the tangent line *T*. Suppose also, the sum of two distances *RA* and *RB* is always the same as the sum of two distances *RA* and *RQ*. So though *R* is moving along the line *T*, *B* and *Q* are always in a circle centered at *R*.

Fig. 0

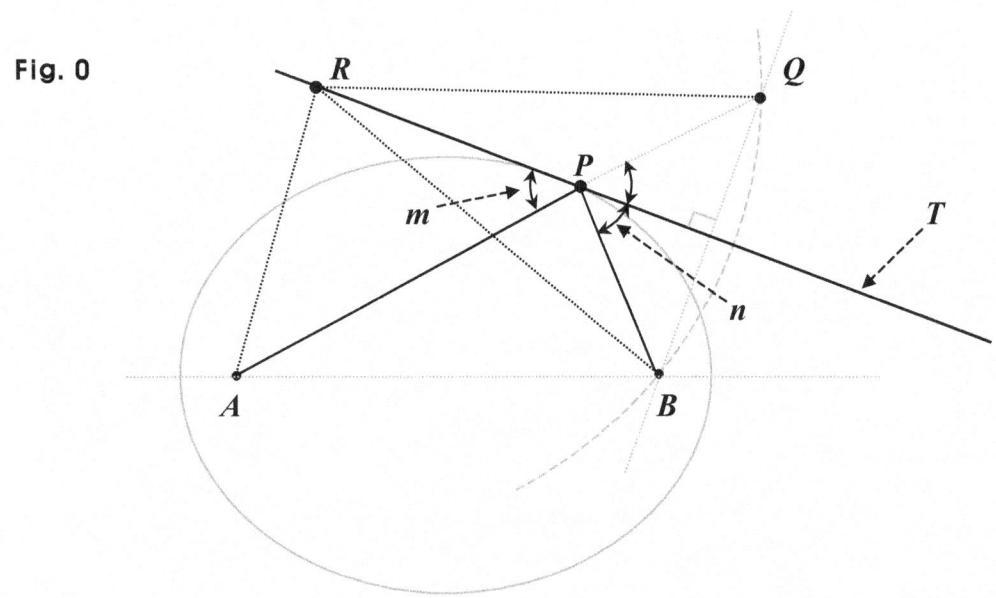

Then, *A*, *P*, and *Q* are in a line, and *B* and *Q* are symmetric about the tangent line *T*, which is thus, perpendicular to *BQ*, and bisects the angle *BPQ*. So the triangle *BPQ* is an isosceles triangle.

Next, the angle *m* and the angle between *PQ* and *T* are opposite angles, so both angles are the same. And we know that the angle *n* is the same as the angle between *PQ* and *T*. So the angle *m* is the same as the angle *n*. The solution above is quite succinct, and is said to have been produced by Akiyama Hitoshi, a Japanese mathematician.

And if *m* is called an angle of incidence, *n* can be called the angle of refraction. So in an ellipse, the angle of inclination is the same as the angle of refraction.

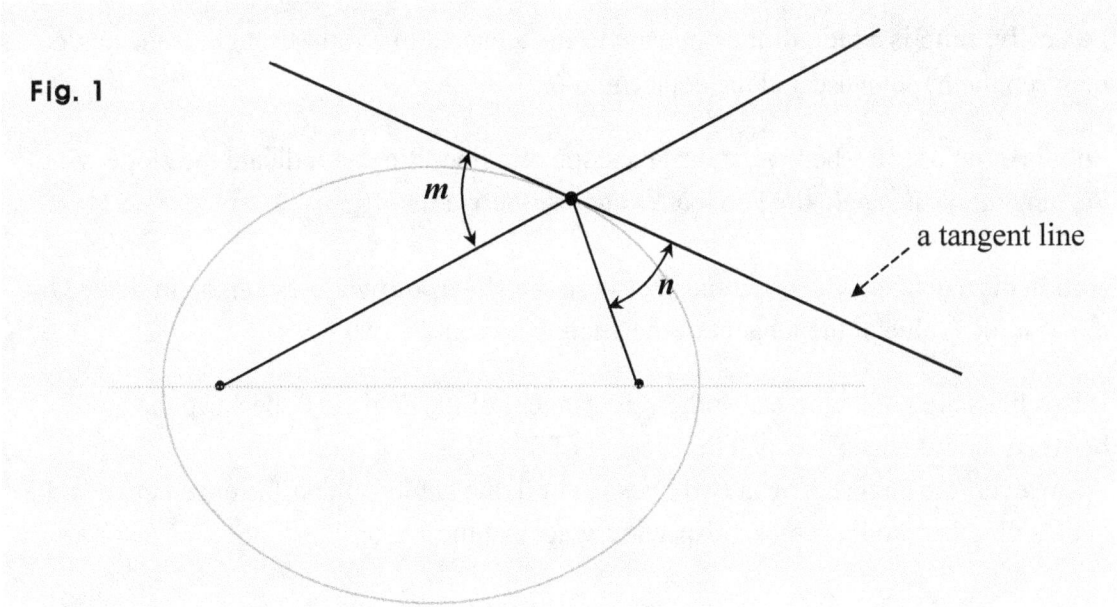

Fig. 1

a tangent line

So an ellipse has a physical property as follows.

If for instance, a light beam comes from one focus, and reflects off the ellipse, it goes to the other focus. So all the light beams coming from one focus and reflecting off the ellipse gather at the other focus.

So we can use such a property constructing a reflecting telescope.

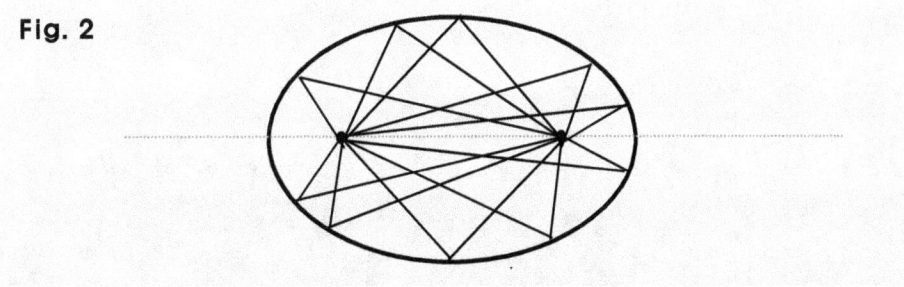

Fig. 2

And also, there can be different ways to the solution, of course. If you like complex calculations, you might like to refer to the method as follows. The method uses a trig-ratio, **tan**, and uses trigonometric identities as follows:

$$\tan(A+B) = \frac{\tan A + \tan B}{1 - \tan A \tan B}, \quad \tan(A-B) = \frac{\tan A - \tan B}{1 + \tan A \tan B}, \quad \text{and} \quad \tan(\pi - \theta) = -\tan\theta.$$

Basically, **tanθ** is a ratio of the opposite to the adjacent in a right triangle if the angle between the hypotenuse and the adjacent is *θ.*

So it is quite useful when we indicate a slope of a line. We can indicate the slope with the tangent of the angle the line makes against the *x*-axis.

And in the *x-y* plane, the difference between the slopes of two lines can be indicated by the absolute value of the tangent of the angle between the two lines.

If two lines are parallel to each other, the tangent of the angle is 0, because the angle between the two lines is 0, and the tangent of 0 is 0.
If however, the angle between two lines is not 0, the angle is the difference between the angles that the two lines make respectively against the *x*-axis.

Fig. 2

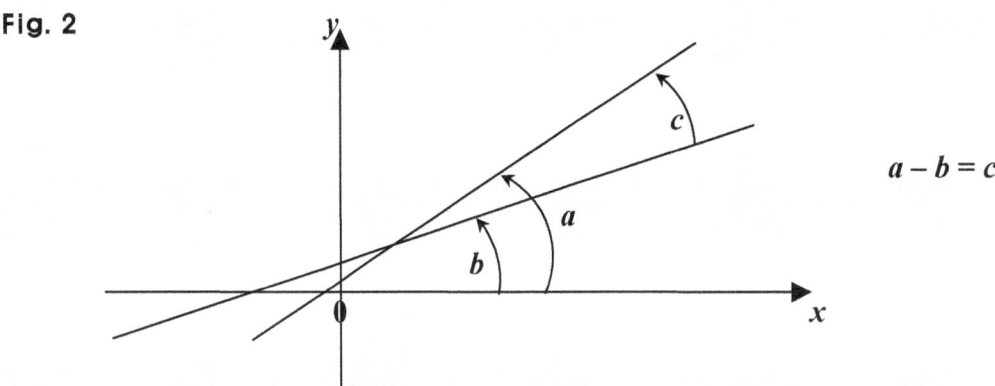

$$a - b = c$$

Fig. 3

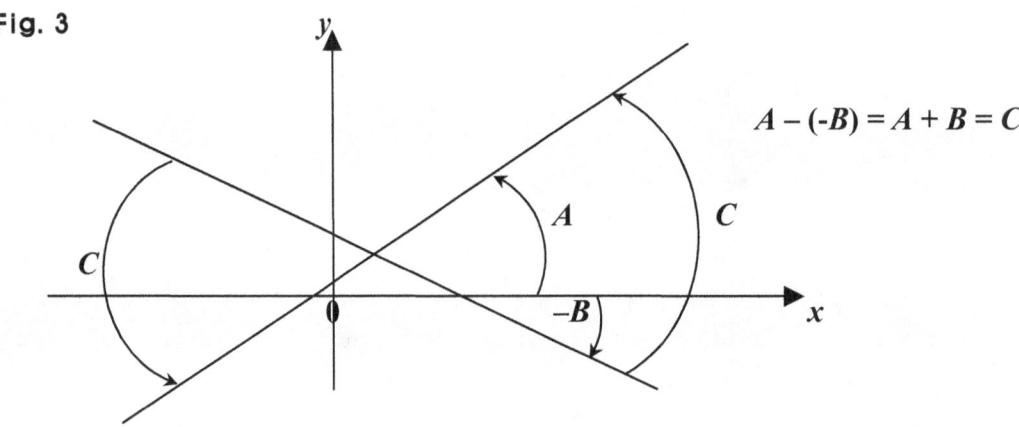

$$A - (-B) = A + B = C$$

Let's see now, how we can get: $m = n$ in this example, using the idea of the ratio, called the tangent.

So suppose first, the ellipse is: $x^2/a^2 + y^2/b^2 = 1$, which is centered at the origin. And assuming next, g is the angle PAB, h is the angle PBA, k is the angle between T and the x-axis, and the two foci are $A(-c, 0)$ and $B(c, 0)$, we can put the problem in a graph the way as follows:

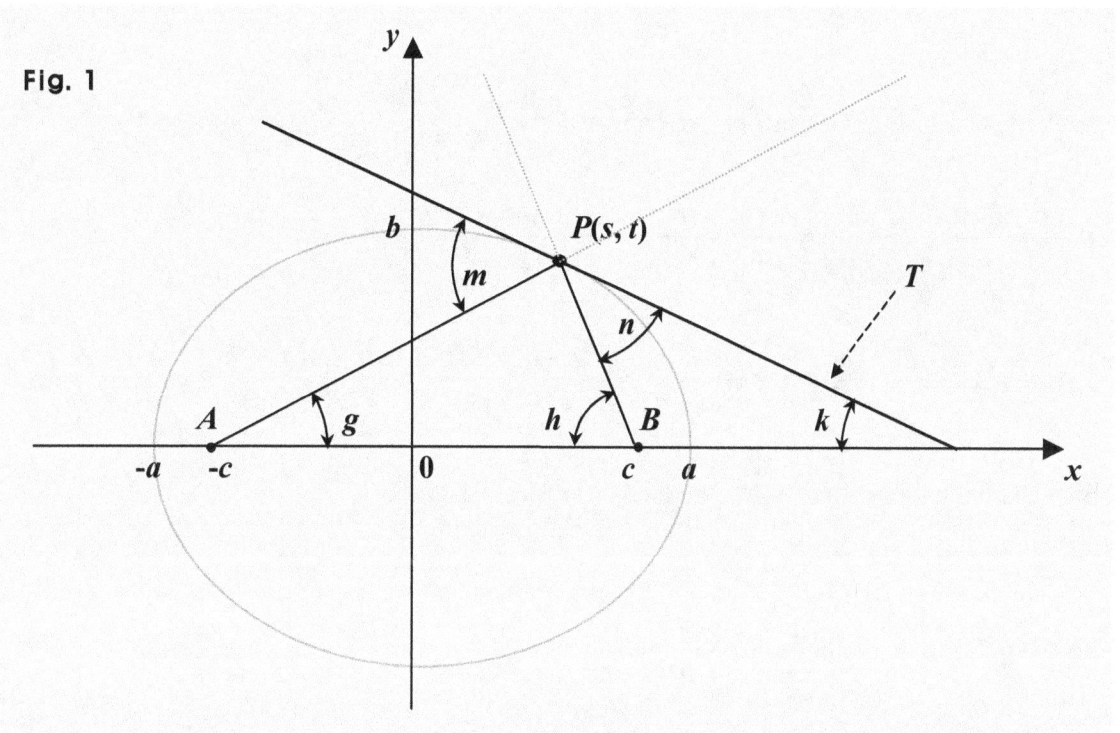

Fig. 1

Then, we get: $\tan m = \tan(g + k)$ and $\tan n = \tan(h - k)$. And by the trigonometric identities, we get:

$$\tan m = \tan(g + k) = \frac{\tan g + \tan k}{1 - \tan g \tan k} \quad \text{and} \quad \tan n = \tan(h - k) = \frac{\tan h - \tan k}{1 + \tan h \tan k}$$

Assuming next, u is the slope of AP, v is that of BP, and w is that of T, we can get:

$$u = \frac{t - 0}{s - (-c)} = \frac{t}{s + c}, \quad v = \frac{t - 0}{s - c} = \frac{t}{s - c}, \quad \text{and} \quad w = -\frac{sb^2}{ta^2}, \text{which was found in the}$$

example in **Examples 7 in Ellipses**.

And we have: $\tan(\pi - \theta) = -\tan\theta$.

So we get: $\tan g = u$, $\tan h = -v$, and $\tan k = -w$.

That's because: $\tan h = -\tan(\pi - h) = -v$, and $\tan k = -\tan(\pi - k) = -w$.

Thus, we get: $\tan m = \dfrac{u - w}{1 + uw}$ and $\tan n = \dfrac{-v + w}{1 + vw}$.

So next, beginning with $\tan m$, we can get first:

$$u - w = \frac{t}{s+c} - (-\frac{sb^2}{ta^2}) = \frac{t}{s+c} + \frac{sb^2}{ta^2} = \frac{t^2 a^2 + sb^2(s+c)}{ta^2(s+c)}.$$

$$1 - uw = 1 + \frac{t}{s+c} \cdot \frac{sb^2}{ta^2} = 1 + \frac{sb^2}{a^2(s+c)} = \frac{a^2 s + a^2 c - sb^2}{a^2(s+c)} = \frac{a^2 c + s(a^2 - b^2)}{a^2(s+c)} = \frac{a^2 c + sc^2}{a^2(s+c)},$$

because we have: $c^2 = a^2 - b^2$, since c is the focal distance.

So we get: $\tan m = \dfrac{u - w}{1 + uw} = \dfrac{t^2 a^2 + sb^2(s+c)}{ta^2(s+c)} \cdot \dfrac{a^2(s+c)}{a^2 c + sc^2} = \dfrac{t^2 a^2 + s^2 b^2 + csb^2}{t(a^2 c + sc^2)}.$

Next, we know that $P(s, t)$ is in the ellipse $x^2/a^2 + y^2/b^2 = 1$.

So we get: $s^2/a^2 + t^2/b^2 = 1 \Rightarrow s^2 b^2 + t^2 a^2 = a^2 b^2 \Rightarrow s^2 b^2 + t^2 a^2 + csb^2 = a^2 b^2 + csb^2$.

Thus, we get: $\tan m = \dfrac{t^2 a^2 + s^2 b^2 + csb^2}{t(a^2 c + sc^2)} = \dfrac{a^2 b^2 + csb^2}{t(a^2 c + sc^2)} = \dfrac{b^2(a^2 + cs)}{ct(a^2 + cs)} = \dfrac{b^2}{ct}.$

So we get: $\tan m = \dfrac{b^2}{ct}$.

And next, moving on to **tan n**, we have: $\tan n = \dfrac{-v+w}{1+vw}$. Then, we can get first:

$$-v+w = -(\frac{t}{s-c}+\frac{sb^2}{ta^2}) = -\frac{t^2a^2+sb^2(s-c)}{ta^2(s-c)}.$$

$$1+vw = 1-\frac{t}{s-c}\cdot\frac{sb^2}{ta^2} = 1-\frac{sb^2}{a^2(s-c)} = \frac{a^2s-a^2c-sb^2}{a^2(s-c)} = \frac{s(a^2-b^2)-a^2c}{a^2(s-c)} = \frac{sc^2-a^2c}{a^2(s-c)},$$

since we have: $c^2 = a^2 - b^2$, where c is the focal distance.

So we get:

$$\tan m = \frac{-v+w}{1+vw} = -\frac{t^2a^2+sb^2(s-c)}{ta^2(s-c)}\cdot\frac{a^2(s-c)}{sc^2-a^2c} = -\frac{t^2a^2+s^2b^2-csb^2}{t(sc^2-a^2c)} = \frac{t^2a^2+s^2b^2-csb^2}{t(a^2c-sc^2)}.$$

Next, we know that $P(s, t)$ is in the ellipse $x^2/a^2 + y^2/b^2 = 1$.

So we get: $s^2/a^2 + t^2/b^2 = 1 \Rightarrow s^2b^2 + t^2a^2 = a^2b^2 \Rightarrow s^2b^2 + t^2a^2 - csb^2 = a^2b^2 - csb^2$.

Thus, we get: $\tan n = \dfrac{t^2a^2+s^2b^2-csb^2}{t(a^2c-sc^2)} = \dfrac{a^2b^2-csb^2}{t(a^2c-sc^2)} = \dfrac{b^2(a^2-cs)}{ct(a^2-cs)} = \dfrac{b^2}{ct}$.

So we get: **tan m = tan n $\Rightarrow$ m = n**.

And in fact, we get: **tan m = tan n = $|b^2/ct|$**, because m and n both are acute angles.

What if however, the tangent point $P(s, t)$ is right above or below the focus?

Then, the line segment connecting the focus and **P** is perpendicular to the **x**-axis, and therefore, does not have a slope.
That's because a vertical line has no slope, because a slope cannot be defined for a vertical line, because a slopes is: rise over run, but in this case, the run is 0, which is the denominator. No denominator can be 0.

So in the cases where $s = \pm c$, we cannot directly apply the calculations above as is. What then, can we do?

In such a case, we can find **tan m** and **tan n** finding the actual values of t when $s = \pm c$, and using the values in the final expression above. So let's now, find first, the values of t.

We know $P(s, t)$ is in the ellipse, and if P is right above or below the foci, we get: $s = \pm c$.

So we get: $x^2/a^2 + y^2/b^2 = 1 \Rightarrow s^2/a^2 + t^2/b^2 = 1 \Rightarrow c^2/a^2 + t^2/b^2 = 1 \Rightarrow c^2 b^2 + t^2 a^2 = a^2 b^2$

$\Rightarrow t^2 a^2 = a^2 b^2 - c^2 b^2 = b^2(a^2 - c^2) = b^4$, since $c^2 = a^2 - b^2 \Rightarrow b^2 = a^2 - c^2$.

So we get: $t^2 a^2 = b^4 \Rightarrow t^2 = b^4/a^2 \Rightarrow t = \pm b^2/a$.

And thus, when P is right above or below the focus, we get: $P(-c, \pm b^2/a)$.

Now, we have: **tan m** = **tan n** = $|b^2/ct|$.   So we get:

$|b^2/(ct)| = |b^2/(\pm cb^2/a)| = |1/(\pm c/a)| = |\pm a/c| = |a/c| = a/c$, since a and c are lengths.

Therefore, if P is $P(\pm c, \pm b^2/a)$, we get: **tan m** = **tan n** = a/c.

And in fact, the value above is the reciprocal of the eccentricity, because we have: $e = c/a$, where e is the eccentricity.